Cambridge Primary
Revise for Primary Checkpoint

Science

Rosemary Feasey
and **Andrea Mapplebeck**

Teacher's Handbook

HODDER EDUCATION
AN HACHETTE UK COMPANY

Acknowledgements
The Publishers would like to thank the following for permission to reproduce copyright material. Every effort has been made to trace or contact all copyright holders, but if any have been inadvertently overlooked the Publishers will be pleased to make the necessary arrangements at the first opportunity.

Photo acknowledgements
p. 7 *cc* © Jane Vasileva/Adobe Stock Photo; **p. 8** *cc* © Volod 2943/Adobe Stock Photo; **p. 38** *br* © J Stock/Adobe Stock Photo.

t = top, *b* = bottom, *l* = left, *r* = right, *c* = centre

Orders: please contact Hachette UK Distribution, Hely Hutchinson Centre, Milton Road, Didcot, Oxfordshire, OX11 7HH. Telephone: +44 (0)1235 827827. Email education@hachette.co.uk. Lines are open from 9 a.m. to 5 p.m., Monday to Saturday, with a 24-hour message answering service. You can also order through our website: www.hoddereducation.com

This edition published in 2022 by
Hodder Education
An Hachette UK Company
Carmelite House
50 Victoria Embankment
London EC4Y 0DZ

www.hoddereducation.com

The authorised representative in the EEA is Hachette Ireland, 8 Castlecourt Centre, Dublin 15, D15 XTP3, Ireland (email: info@hbgi.ie)

Impression number 10 9 8 7 6 5 4 3 2
Year 2026 2025 2024 2023

Cover illustration by Lisa Hunt, The Bright Agency

Illustrations by Natalie and Tamsin Hinrichsen, Vian Oelofsen, Stéphan Theron, James Hearne

Typeset in FS Albert 11/13 by IO Publishing CC

Printed and bound by CPI Group (UK) Ltd, Croydon, CR0 4YY

A catalogue record for this title is available from the British Library.

ISBN 9781398343115

FSC www.fsc.org MIX Paper | Supporting responsible forestry FSC™ C104740

Contents

Introduction

What is the Study Guide about?

The aim of the Study Guide is to help learners to recall key information and ideas and develop their understanding about the science topics that they have been learning throughout Stage 6. It will help them to make sure that what they learn stays in their memory for a long time.

Learners can work through the Study Guide on their own or with a partner at school or at home, or both. Space is provided for learners to write their answers but the Study Guide also contains questions which demand that they communicate their revision in different and creative ways such as producing posters, mapping out their ideas on large sheets of paper and even making things, all of which are aimed at helping learners remember and deepen their understanding of the science that they have been learning.

Why is revision important?

Revision gives learners an opportunity to reflect on what they have learnt and remembered. It helps them to identify what they know, what they do not know and what they are still unsure about. Revision of course helps learners to develop learning in different topics, but it also helps to build confidence, not just in what they know, but developing their ability to use ways to access what they have learnt. By revising topics, learning is extended and deepened, meaning that learning is more thorough and so individuals are more able to apply their learning to new contexts and of course in the future to tests and exams.

Revising an area of science helps learners to access knowledge and understanding, at the beginning this can be quite superficial, but by accessing memory, learners are able to retrieve more and more as their brains make links and store information in the long-term memory.

What is important is revision should not be tedious; the aim of the Study Guide is to use approaches that will interest and engage learners to review their learning.

How can the Study Guide help learners?

Revision activities aim to support learners to remember facts, information, ways of working, and to make connections between ideas within a topic and across scientific areas of learning. In this way, the knowledge and understanding they develop is more likely to stay in their memory for longer. Engaging learners in revision activities is of course one way of eliciting what they know, identifying misconceptions and finding out how deep their understanding is are other ways too.

How can an adult help learners when using the Study Guide?

Encourage learners to be honest with themselves, discuss that if they do not know something it is important to admit it otherwise they cannot be helped to understand an idea or carry out a task. Equally if they think they know something but are unsure they should also acknowledge that and seek help or clarification. Sometimes, even after support, a learner may still not understand an idea or a word, in which case you might need to go back a few steps in their learning and/or help them to unpick what it is they do not understand. In this way, you can produce bespoke explanations or examples to help move their learning forward.

Revision should be seen as a collaboration between learners and the teacher. Reinforce the idea that it is okay to make mistakes and errors, especially when revising. Revision is about finding out what we know, are unsure about and what we either have forgotten or did not understand in the first place. Encourage learners to use their notebooks, as well as the *Cambridge Primary Science Stage 6 Learner's Book* to help remind them about what they have learnt.

When working through the Study Guide with learners, frequently ask them how they knew the right answer, just as you do in Mathematics, so that learners engage in metacognitive thinking (thinking about their thinking). In doing this, learners will articulate their thinking and you will be able to access the depth of their understanding through explanations and the scientific language that they use.

Learners should understand why strategies are useful to them, so that they can choose which approach to use in science, as well as in other subject areas.

Revision approaches

The Study Guide uses different approaches to revision and provides an explanation of them so that learners understand how they support the development of their understanding of science ideas and embed learning in memory. Different people learn and remember their science in different ways, so using a range of approaches helps to individualise revision. It is important that learners understand which approaches they find the most helpful and use them frequently to support their learning not just in science but across different subjects.

Some of the revision approaches used in the Study Guide are explained below, so that you know what they are, why they are used and how they can support learners' understanding.

Concept maps

A concept map is a type of memory map to help learners remember key words and ideas in a topic and make links between them.

With a concept map, learners map out key words and ideas and make links between them, displaying their understanding by drawing lines between ideas or words and then writing on the lines to explain the connections. It is the connections between words/ideas that are the most important aspect of a concept map because the learner can access their deeper understanding.

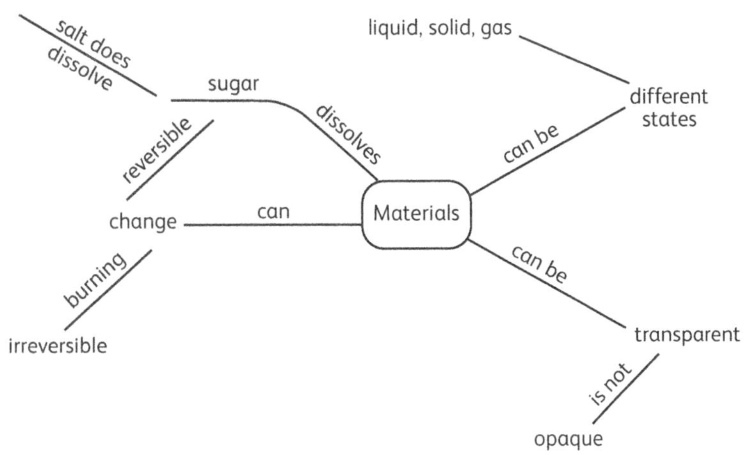

Definitions

Thinking about and writing down definitions for words can help learners to understand them better and give them confidence to use the words when speaking or writing about science.

Being able to use scientific vocabulary correctly is very important, so ensure that learners know how to use the glossary at the back of the *Cambridge Primary Science Stage 6 Learner's Book*, or dictionary and thesaurus to check if they are correct. Encourage learners to make their own personal science dictionary, writing the word, how to say it, a definition and even drawing or sticking a picture or sample of something, for example, a material (aluminium foil) to show understanding and support memory. Key word and fact file cards are two approaches that support learners in not only being able to read and spell scientific vocabulary but also to show that they know what words mean and their ability to give everyday examples and apply them to different contexts.

Double bubble

A double bubble is another type of memory map to access memory and assess if learners have understood key ideas and can make connections between them. The double bubble helps learners to look at two different ideas and compare them for similarities and differences, contrasting areas of learning involving objects, animals, plants, key concepts, events, and processes.

In the centre of the large circles learners write the name of, for example, the two ideas that they are comparing. Down the middle of the double bubble in the hexagons they note similarities. This means they will write one thing in each hexagon that is true for both ideas. In the outside circles they then consider differences, these circles are coloured so that ideas in the same circle are connected. Learners will note things that are different for each of the ideas they are exploring.

Drawing out their thinking using this type of memory map encourages learners to make connections and remember what they have been learning about.

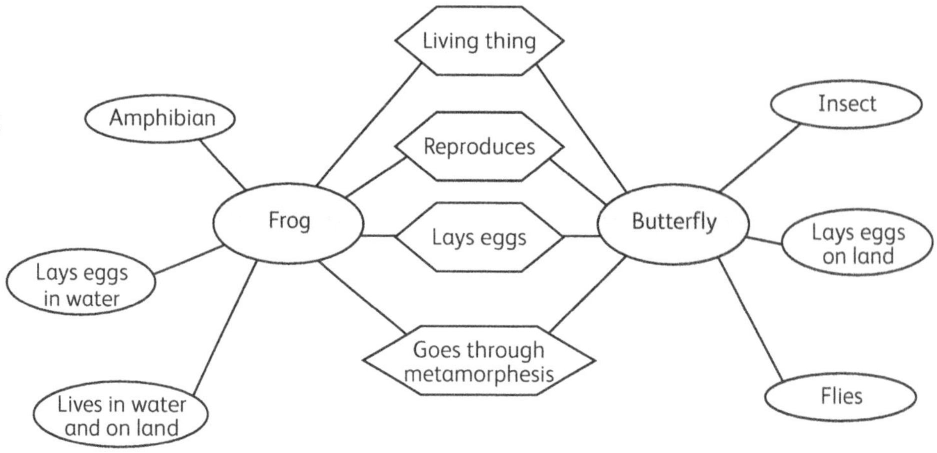

Frayer map

A Frayer map, is another thinking map that helps learners to identify and define concepts that they are familiar and unfamiliar with, related to key science vocabulary. Using a Frayer map supports the learning of vocabulary that is needed for learners to talk and write about their science. Using this approach scaffolds learning how to spell words and understand key scientific words so that they can read, spell, and explain scientific words and finally apply them to everyday life, the latter is evidence for formative assessment that learning is secure.

A Frayer map is used to focus on revising and learning about one word at a time.

Definition	**Characteristics/features/facts**
A Moon that goes around (orbits a planet or a star). A machine that goes around (orbits) the Earth.	Smaller than the planet or star it orbits. Sends information back to Earth.
Examples	**Non-examples**
The Earth's Moon Titan is a satellite of Saturn.	The Sun meteorites a bus

Fishbone organiser

This is another type of thinking map. A fishbone organiser is a useful way of organising ideas and knowledge related to the same topic. It is a visual way to link many ideas that relate to the same issue and help record what learners know about a problem and how to solve it. The aim of a fishbone organiser is to help organise ideas in a simple way that helps learners to make links and solve problems. Encourage learners to use different colours on their fishbone organiser so they can pick out ideas that link.

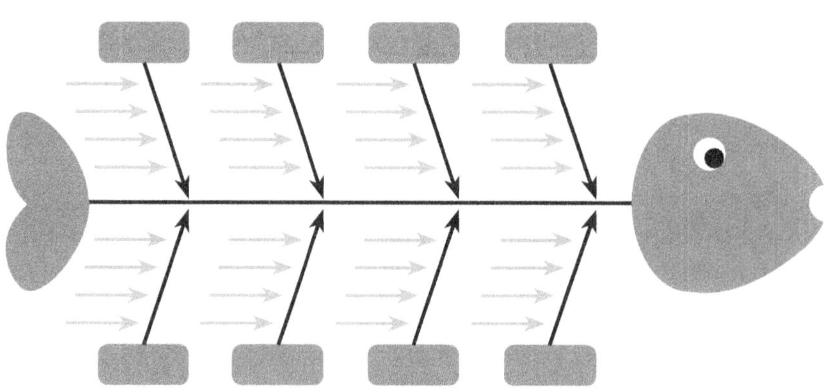

Infographics

The word infographic splits into two parts:

info (information) + **graphic** (visual)

An infographic is made using pictures, charts and graphs. It is a visual organiser designed so that information can be read easily.

Using infographics can help the brain remember information, as it draws on the idea of Dual Coding. 'Dual Coding' provides two different representations of the information, both visual and verbal at the same time. The idea is that learners look at the information on an infographic and also talk about it with others, this approach can make it easier for the brain to recall (remember) information.

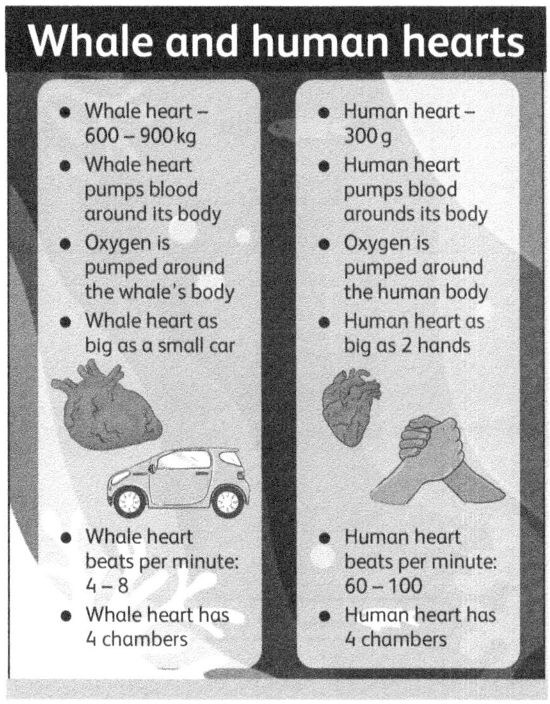

Key word cards

The purpose of key word cards is to support learners in remembering words and to revise key scientific vocabulary so that they are confident in being able to read, spell, and know what the word means. Key word cards can be created and used in any topic and are frequently used throughout the Study Guide. On one side of the card learners write the key scientific word, on the other side the definition. To aid memory retention, learners could write words in different colours and split words up to help memorise them. While making the cards can be beneficial, it is the act of using the cards throughout a topic which will help learners to retain scientific vocabulary and embed it in long-term memory. Learners should use the cards to test each other, create a second set of cards and play pairs. The cards are turned upside down and they have to match pairs, for example, pairs of words, definitions or words to match definitions.

There must be two objects that touch. These can change the motion, direction and shape of the object they act on. They can be balanced or unbalanced.

Contact force

There must be two objects that interact, these objects are not touching. Examples of non-contact forces are weight (gravity) and magnetism. They can be balanced or unbalanced.

Non-contact force

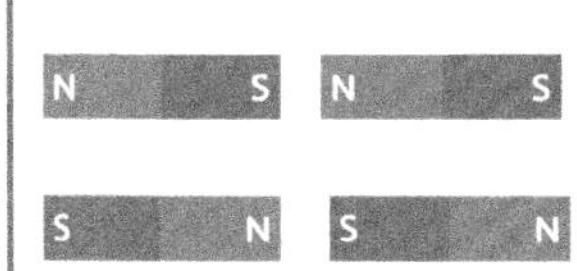

Fact files

Fact files are another useful approach to revision to support learners in learning, memorising and recalling key information. They are quick and easy to make and each card has some or all of the following:

- **i** key idea/concept
- **ii** linked scientific vocabulary
- **iii** examples of scientific idea/concept in everyday life.

It is the final point that provides an excellent formative assessment point for the teacher; where learners are able to apply an idea/vocabulary to a new context the learner is demonstrating whether learning is or is not secure.

As learners make the different fact files for a topic or different topics make sure that they keep them and use them to revise learning either themselves or giving them to friends and family to 'test' their science.

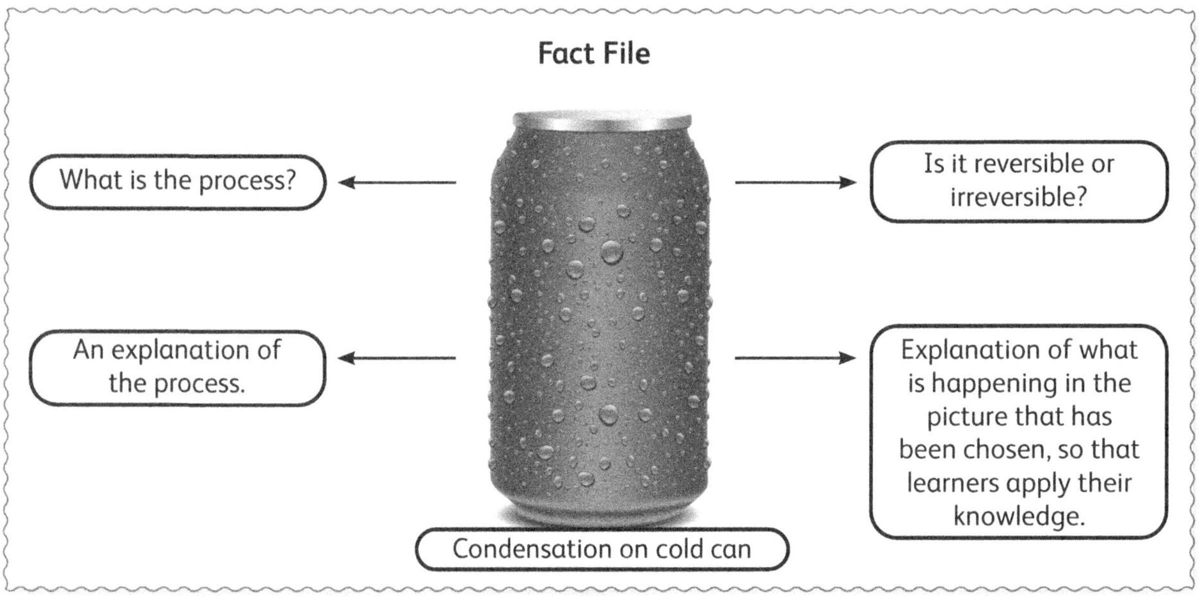

Learning flower

A learning flower is another memory aid to support learners in remembering and organising ideas and vocabulary. Visual approaches to revision such as this help learners to recall key ideas and organise them in such a way that they can make links between the different knowledge.

For example, learners could organise their revision of a topic by placing:

1. the topic title in the centre of the flower
2. key words on the front of the petals
3. definitions of the words on the back of the petals
4. key learning on the stem
5. what they already knew prior to the topic on the roots
6. questions or things they do not fully understand on the leaves.

Memory maps

A memory map is a way for learners to organise information that they know. For example: Materials is a broad topic, it would help learners to organise learning into sub-areas: Properties of materials, Names of materials, Dissolving, Changes in state, Reversible and irreversible changes. Organising ideas in this way can help learners to retrieve connected information, as well as show what they know and where the gaps are, for example, if a learner writes very little under the heading Dissolving.

A memory map differs from a concept map, the latter requires that learners show the links between words and ideas using words or phrases written on the lines linking words.

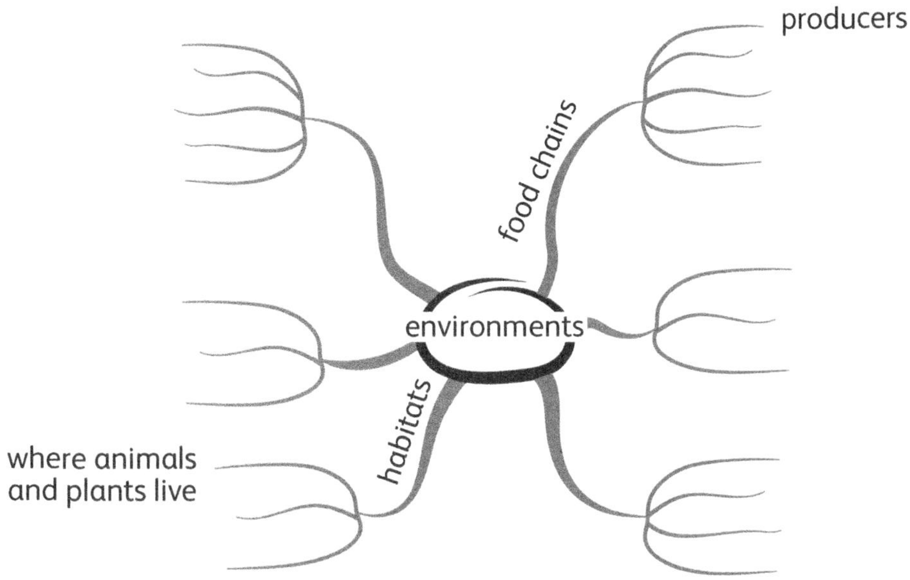

Mnemonics

Mnemonics often use rhymes or a sentence to help learners to memorise information. As with all approaches seeking to support learners to retain and recall information, the aim is to develop 'sticky memory' so that learning literally sticks in their memory making it easier to retrieve information and ideas.

Mnemonics are useful not only to help learners remember a set of information, for example, the names of the planets in our solar system but they can also add another dimension such as the order of the planets as they orbit the Sun. By creating and using mnemonics learners are able to access one or more pieces of information from their memory.

Mnemonic for the planets:

My (Mercury) Very (Venus) Easy (Earth) Method (Mars) Just (Jupiter) Speeds (Saturn) Up (Uranus) Nothing (Neptune).

Model answer (Worked answer)

A model/worked answer can be an 'ideal' response to a question, or it could be one where there are errors that have been purposefully included to identify whether learners have the same difficulties. Looking at different answers to questions and thinking about how they can be improved demands that learners use their personal knowledge to make sense of an answer. Challenging learners to read and evaluate an answer demands that they access their personal knowledge in science and recall ideas from their memory, thus revising their science. By analysing the strengths and weaknesses of an answer, learners are engaging in a form of peer assessment. Requiring learners to edit or rewrite and improve an answer ensures that they apply learning and consider what constitutes a good answer. To further support learners, you could engage them in a discussion about the criteria for a good answer on a specific topic, and learners then write to those criteria.

Prefixes

A prefix is a group of letters added to the beginning of a word to change the meaning of the word. Teaching prefixes (and suffixes) in science helps learners to de-code words and understand their meaning. This is important because it helps learners to understand that some words in science can be broken down into parts which in turn helps them remember words and their meaning.

For example, the word 'microbe' begins with the root word **micro** which comes from the Greek word **mikros** which means 'small', whenever you come across a word where the root word is 'micro' the word has something to do with being small.

The word **microscope** is made up of the root word **micro** and the word **scope**. 'Scope' means to see, 'micro' means small. So, the prefix changes the meaning of 'to see', to 'to see very small (microscopic) things'.

Rich picture poster

A rich picture is a way of showing an idea, information, a process or for example, a habitat by using pictures, diagrams and individual words, phrases, and colour coding. Using a rich picture can sometimes be easier for learners to communicate what and how much they know than, for example, writing sentences or paragraphs, especially if some individuals find extended writing challenging.

The content of a rich picture does not have to be created in a specific order, it can just show the flow of ideas about, for example, how to solve a problem such as how to encourage people to recycle and reuse materials or a set of ideas linked to friction. A rich picture differs from an infographic because it does not have to be based on communicating data using graphs and charts alongside words and diagrams.

Revision hexagons

Using revision hexagons is a challenging approach as many connections can be made between ideas. Hexagons are specifically used because they tesselate, fit together with no gaps. The tessellated hexagons are used to structure thinking, the challenge for learners is to make sure that whatever words, facts, or ideas are placed in a hexagon, they must be able to articulate the links it makes to the other hexagons bordering on each side. This idea of linking is crucial to revision hexagons.

This revision approach helps learners to:

- think about what they know
- recall facts and ideas
- consolidate ideas
- make links between learning
- indicate depth and breadth of learning.

While most learners will write text into hexagons others could draw or cut and paste pictures into the hexagons.

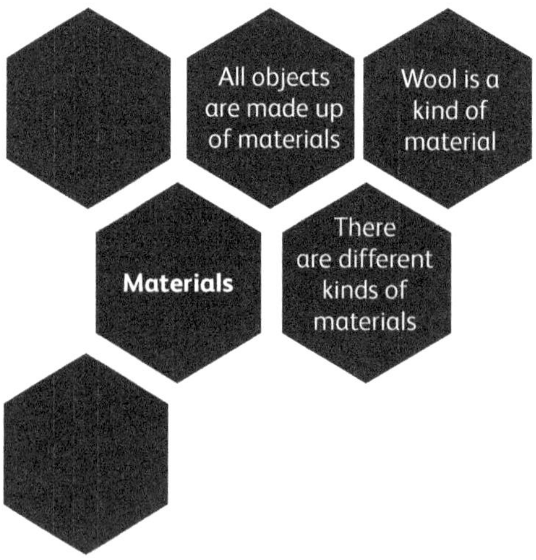

Venn diagrams

A Venn diagram is another thinking map specifically designed to challenge learners to systematically arrange learning to show the relationships between items, similarities and differences.

For example, learners could use a Venn diagram to organise objects according to whether they are thermal conductors or insulators. Venn diagrams usually have 2 or 3 circles which overlap in the middle; where the circles overlap this shows similarities, and where they do not overlap shows differences.

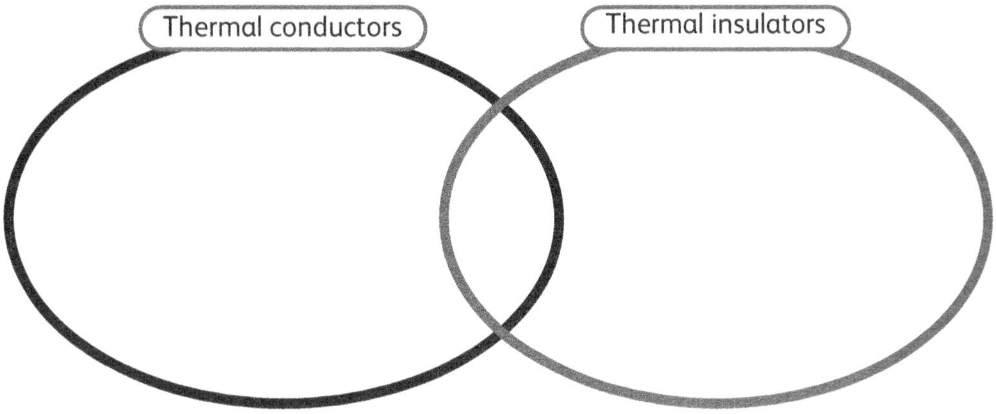

Some activities have associated templates which can be downloaded and printed out for learners. These templates are available on www.hoddereducation.co.uk/cambridgeextras

Learners using these different approaches in their learning and revision:

As learners work through their Study Guide encourage them to personalise their learning by challenging them to think about which approaches they find the most useful to:

- access their memory and remember ideas and words
- make connections (see links) between learning
- remember the most ideas or information
- organise what they know
- identify gaps in their learning and understanding.

The final point is important for learners to consider, using approaches that they enjoy can make revising a topic more interesting, easier to learn and remember, which of course is the purpose of revision. By engaging learners in thinking about their own learning and identifying which approaches are most suited to them as individuals, you are helping them to identify bespoke approaches they can draw on throughout their lives as learners.

Unit 1 Systems and diseases

The circulatory system

Study Guide pages 12–16

Learner's Book pages 9–14

Objectives

Learners will revise:

- many vertebrates have similar circulatory systems
- how the heart works
- the function of the heart.

Science background information

It is important that learners are given time to reflect on what they already know about how the circulatory and respiratory systems work. Encourage them to refer to the Learner's Book and any exercise books that they might have worked in.

Listen to learners discussing their ideas and read their memory maps, checking to see if they have remembered ideas correctly, what they have missed out and any misconceptions. As you work with individuals or groups, note common omissions which might indicate areas where learners are unsure, or learning is not embedded.

A common misconception about the heart and circulatory system is that the circulatory system has a 'single loop'. Learners may think that the arteries carry blood from the heart to the body and the veins carry blood back to the heart from the body. In this misconception, learners are 'missing' the role of the lungs in the circulatory system.

Display a picture of the circulatory system and talk through the idea that blood is pumped from the left side of the heart to the body, it then returns to the right side of the heart. From there blood is pumped to the lungs, where carbon dioxide is exchanged for oxygen. This blood then returns to the left side of the heart, where it is pumped around the body.

Humans are vertebrates, so are monkeys, elephants, and mice, they all have similar circulatory and respiratory systems. The heart's main role is to make sure that blood travels around the body distributing oxygen and nutrients. Blood is pumped by the heart through the aorta to veins which carry the blood to capillaries which deliver the oxygen and nutrients to cells around the body. This system is circular, it has no end, it goes round and round. The returning blood goes to the lungs where it picks up oxygen and returns to the heart where the journey is repeated, all day and night. Most mammals, including humans, have this type of circulatory system.

Key to understanding, is that this system is closed, which means that blood is continually moving around human bodies and other vertebrates. In this activity, the circulatory systems of humans and whales are compared. Whales are vertebrates like humans, some learners might think, incorrectly, that they are fish because they live in water.

Revision approach background information

A memory map is a way of helping learners to think through and explore what they know about a topic. Learners can write words, phrases and draw symbols or pictures about facts and ideas and use lines to make links between different learning on their map. Key ideas can be placed in larger boxes, bubbles, or a specific colour, with other ideas and facts in smaller text or a different colour, showing a hierarchy in their ideas.

The benefit of using memory maps is the way in which the memory is stimulated, with one idea leading to another. The learner is able to build up different layers of information as the memory is prompted to remember more.

Memory maps can be changed, either because the learner recognises that what they have included needs to be amended or new learning is added. This means that memory maps should be seen as fluid pieces of work which can be modified at different points, including at the end of revision as a self-check.

Memory maps can be created by individual learners or by a group of learners who share ideas and facts, reminding each other of learning and sometimes sharing new ideas.

Some learners benefit from key vocabulary being displayed which helps to prompt their memories to locate information and ideas associated with those words.

Infographics are visual representations of information and data which allow the learner to access information quickly and easily. Infographics use visual communication, with less emphasis on text and more on pictures, charts, diagrams, and graphs. Infographics can capture the attention of learners, particularly those whose learning style is more visual.

Starter activity suggestions

- With learners, create a memory map related to a random topic which all learners will know something about, for example, a sport or an animal. Discuss how to construct the memory map and why it is useful for revising learning, for example prompting memory, free flow of ideas, can be added to and amended at any time. Then ask learners to construct their own memory map either working as individuals or in a group. Make sure that they keep their memory map and encourage them to amend and add to it at any point during this revision topic; using different colour pens helps them to compare before and after ideas.
- Alternatively, in small groups, learners draw around a member of the group on a large sheet of paper. In the body outline learners draw how they think blood is circulated around the body. This will help to elicit misconceptions about whether they see the system as a single or double loop.
- Ask learners to discuss why whales are classified as mammals, focusing on the following criteria:
    - they give birth to live young
    - they have hair (although it is very sparse on their body)
    - they have lungs and breath air
    - they provide milk for their young.
- Challenge learners to look at and read the infographic on page 16 of the Study Guide for 3 minutes, concentrating on remembering information. Learners should use strategies such as forming pictures in their minds and remembering words and numbers, for example, heart beats per minute: human 60, whale 4.

Activity notes and answers

Pages 12–13
Activity 1

Check that learners understand how to make their memory maps.

Page 13
Activity 2: Answers

1 and **2** Responses should be similar to those in the table below.

Word	What do they do?
Arteries	Tubes that carry blood away from the heart.
Blood	The fluid in humans and other animals that delivers materials for life to the body's cells.
Blood vessels	Tubes that form a network that carries blood around the body as part of the circulatory system.
Capillaries	Tiny blood vessels that deliver water, oxygen and nutrients to the cells in the body and carry away waste.
Lungs	The internal organs that fill with air when you breathe in.
Veins	Blood vessels that carry blood from the capillaries back to the heart.

3 and **4** Give learners time to check their definitions, ensure that they use the glossary at the back of the *Cambridge Science Stage 6 Learner's Book* or a dictionary and test themselves and each other on how to spell each word correctly and make sure they understand what each word means.

Page 14

Activity 3: Answers

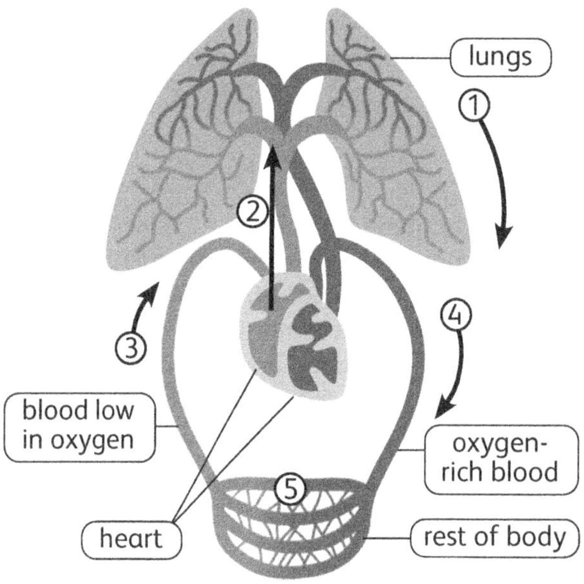

a Check that learners self-assess their own diagram; they could revisit page 10 of the *Cambridge Science Stage 6 Learner's Book.*

b Challenge learners to think about ways of remembering any labels that were incorrect, for example, sketch the diagram on a mini whiteboard or draw it in the air and talk it through.

c When they repeat the activity, use their idea from b) to help them to remember.

d For example:

The circulatory system makes sure that oxygen, water and nutrients are delivered to all parts of the body. The heart pumps the oxygen low blood to the lungs, which exhale carbon dioxide through the nose or mouth. When air is inhaled, oxygen rich blood is sent to the heart, which pumps the oxygen rich blood to the rest of the body.

Activity 4

At various points during this unit ask learners to revisit their memory map and add to or amend as appropriate, this could be done in a different colour pen to show progression in their learning. Discuss their additions and amendments and ask them to explain any new links they have made to check their understanding.

Page 15

Activity 5: Answers

For example:

i. The parts of the heart

ii. How the heart works

iii. How the circulatory system works

iv. How to take the diseased heart out

v. How to put the new heart in the body

vi. How to keep the person alive while the hearts are being swapped.

Pages 15–16
Activity 6: Answers

For example:

Similarities and differences between human heart and whale heart	
Similarities	**Differences**
Heart pumps blood around the body	Whale heart is weighed in kg not g
Oxygen is pumped around their bodies	Whale heart size of small car, human size of 2 hands
Both hearts have 4 chambers	A whale heartbeat is very slow, humans is faster per minute

Page 16
Activity 7

Check that learners choose facts that are similar.

Cross curricular

- Memory maps provide a focus on learners' ability to remember and apply key scientific vocabulary and the ability to use phrases and paraphrase ideas rather than write whole sentences.
- Learning scientific vocabulary is key to learners' understanding and being able to articulate their learning. Learners who struggle with the vocabulary, will of course, find it more difficult to communicate their ideas and understanding and less likely to attain at the appropriate level. Focusing on key vocabulary not only supports learners but also helps teachers to access and support any misconceptions that learners might hold. Encourage learners to use approaches that 'suit' them to learn key scientific vocabulary, for example:
  - break down polysyllabic words, for example: res – pi – ra – tion
  - family and friends testing them
  - mini dictionaries
  - hand signs
  - flip key word cards, choose a card, spell, and define word, then flip card over to check
  - highlight the part of the word that they find tricky. For example, diaphragm and focus on learning the part they find hard to remember.

Further activities

- Encourage learners to share their memory maps with others, so that they can:
  - compare their ideas
  - check and correct their own maps
  - add new information from someone else's map to their own to extend their learning.
- In those minutes between activities, or just before breaks or lunchtimes, engage learners in quick fire word activities such as spelling the word 'diaphragm', or engaging in the game 'I say…' where they have to do what you say quickly, for example, 'I say inhale', so the whole class inhales.

ICT links

- Learners use the internet to search for additional information on the heart, or to watch video clips which explain how the heart works. This can help embed ideas, or learners can check their own ideas against scientifically acceptable explanations and, where appropriate, amend.
- Learners use an online dictionary to check spelling and definitions. Research additional information on, for example, Christiaan Barnard.

The respiratory system

Study Guide pages 17–19

Learner's Book pages 15–18

Objectives

Learners will revise:

- many vertebrates have similar respiratory systems
- how the lungs work
- the function of the lungs.

Science background information

This section focuses on breathing, and challenges learners to remember and apply knowledge relating to the respiratory system in humans. Breathing is when we inhale (breathe in) and exhale (breathe out), moving gases into and out of the lungs. When we inhale either through our mouths or noses, the diaphragm (which is a muscle) contracts and moves downwards, allowing the lungs to inflate. When we exhale, the diaphragm moves upwards and air is forced out of the lungs, at this stage the gas from the lungs is carbon dioxide rich and oxygen poor.

When we inhale, the air moves down the trachea, more commonly known as the windpipe, towards the lungs. At the bottom of the trachea the tube splits into two, the right and left bronchus, each one leads to the lungs. These split again into tiny bronchioles and at the ends of these are air sacs called alveoli, it is here that oxygen from the air breathed in moves into the blood and carbon dioxide from the blood moves out, and the process is reversed, and carbon dioxide is exhaled.

All the parts of the body that enable oxygen to be circulated to the different parts of the human body and carbon dioxide to be removed, is called the respiratory system. It is useful to check that learners know this. Check that learners are not referring to breathing as respiration, this is incorrect, breathing is the mechanical process of inhaling and exhaling air. Respiration takes place at cell level; when we inhale air, it contains oxygen, this is carried from the lungs by the blood to the cells. In cells, oxygen helps to break down glucose and release energy which is needed by each cell so that they can do their 'job' in the body.

Revision approach background information

In this set of activities, learners are introduced to a model that is used to help explain how the body inhales and exhales. Using physical models in science can be useful, particularly for those learners who prefer to handle or see a visual which helps them to understand an idea. Models can slow down a learner's thinking and help them work through a process step by step. Encourage learners to articulate their thinking as they observe or use a model so that their thinking can be heard allowing learning to be reinforced or misconceptions identified and challenged.

Starter activity suggestions

Make sure that each group has a model of the lungs so that they can use it and observe how it works. Learners should break down the process of breathing into small steps, using correct terminology. Learners could video themselves using and explaining the model.

Activity notes and answers

Page 17
Activity 1: Answers

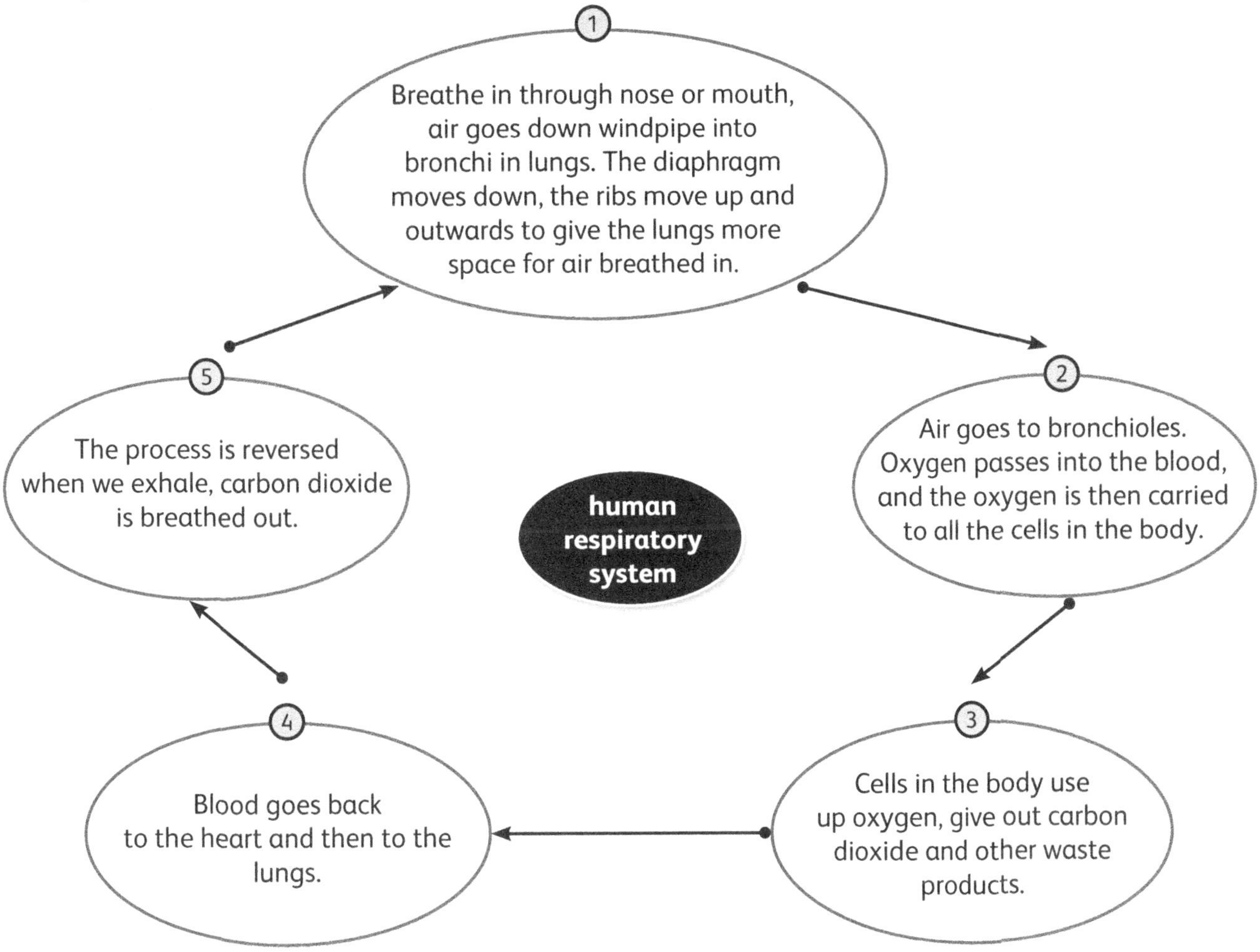

Page 18

Activity 2: Answers

respiratory system	oxygen rich air	carbon dioxide	waste gas
lungs	internal organ	inhale(d)	exhale(d)
breathing	diaphragm	ribs	air
trachea	bronchi	bronchioles	oxygen
cells	respiration	gas exchange	expiration, or exhalation
nose	mouth	heart	

1–3 Discuss with learners which words they have chosen and give them time to write definitions and learn words highlighted in red and orange. Their responses will be individual depending on which words they know; they should write a definition for all the words they have highlighted red and learn how to spell them.

Pages 18–19

Activity 3: Answers

1 The balloon at the bottom of the model is meant to be the diaphragm. It works like a human diaphragm. When the model diaphragm is pushed up it is like the diaphragm in the body, the balloons deflate – this is a person exhaling. When the model diaphragm is pulled down, this represents the diaphragm moving down in the human body, the balloons inflate just like lungs inflate when a person inhales.

3 Learners would not develop an incorrect idea because the model shows one tube, the trachea, which then splits into two, each tube going into one of the lungs.

Page 19
Activity 4: Answers

1 ***Breathing*** takes place when a person inhales (breathes in) and exhales (breathes out).
Respiration is different to breathing because it takes place in the cells of the body.

2 Check learners' additions and changes to their respiratory system memory maps, check that these are scientifically appropriate and ask learners to explain any ideas where you think they might require further support.

Cross curricular

Encourage learners to articulate their thinking. An important element of developing ideas and understanding key scientific language is for learners to have the opportunity to articulate their ideas. In doing this the learner and others can listen to ideas and hear inconsistencies as well as how well an idea/definition is explained. Practising explanations so the learners can 'verbally draft and redraft' sentences supports learning and development of understanding.

Further activities

Learners could use the illustration of the model lungs to create their own model.

ICT links

Learners can video their working model and create a voice-over explanation.

Illness and diseases

Study Guide pages 19–22

Learner's Book pages 19–24

Objective

Learners will revise:

- diseases and their causes and defence mechanisms against infections.

Science background information

Ask learners to list all the different illnesses that they know of and to group them into viruses, bacteria, parasites, or fungi, then check them using the *Cambridge Primary Science Stage 6 Learner's Book* or the internet. During revision, it will be useful to use learners' own experiences, for example, the common cold, flu, or Coronavirus, discussing effects of these illnesses, how they are transmitted and preventative measures.

Revision approach background information

Prefixes and suffixes are very useful for learners to know and understand in science. At different points in the Study Guide, learners will be asked to focus on prefixes and suffixes to help them understand the meaning behind different words. In doing this, learners develop their understanding that unfamiliar words can be broken down and if they know the prefix or suffix they can often work out what the word means themselves.

Activity notes and answers

Pages 19–20
Activity 1: Answers

For example:

i microorganism: very 'small' living creature

ii microbiologist: a scientist who studies 'small' living things

iii microphone: makes a 'small' voice loud

iv micro-habitat: 'small' habitat

Page 20
Activity 2: Answers

1 flu	**B** virus
2 cholera	**A** bacteria
3 athletes' foot	**D** fungus
4 malaria	**C** parasite

Page 21
Activity 3: Answers

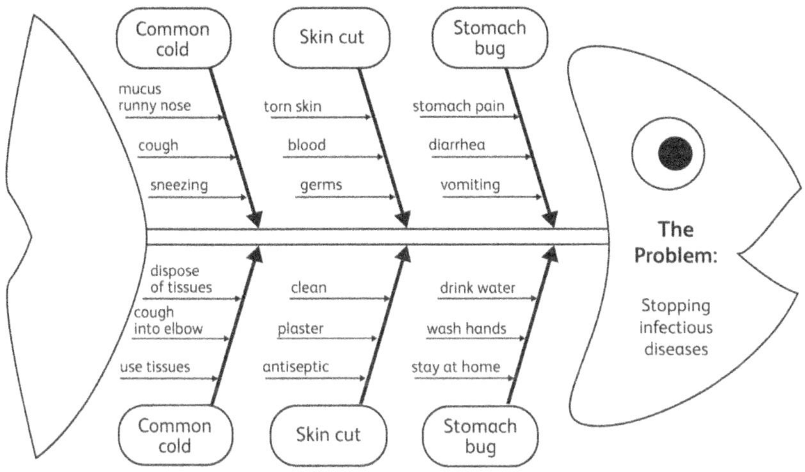

Activity 4

Check that learners add or change their original memory map from the beginning of this unit. Ask them to explain why they have made the changes and to describe how they think their learning has changed.

Activity 5: Answers

Ensure learners read the information in the table and can explain what the results are showing.

1 Accept reasonable answers, for example, Which is the best way to wash hands?

2 Make sure that learners use the data from the table to support their answer. For example:

Scientists were right to tell people to wash their hands for 20 seconds with soap. The results show that only 3 % of the learners' hands were covered with bacteria after washing with soap for 20 seconds. Washing for 20 seconds without soap resulted in 25 % of the hand covered in bacteria. Washing longer with or without soap would be better as we can see that after only 5 seconds with no soap the results were 93 % of the hand covered in bacteria and even with soap only washing for 5 seconds the percentage was high – 86 % of the hand was covered in bacteria. Therefore, washing hands for at least 20 seconds is important.

Pages 21–22
Activity 6

1–4 Check that learners create their plan and share with either another group or partner and discuss how it could be improved and why.

Cross curricular

Learners' ability to transfer and apply dictionary and glossary skills from literacy is important. In these activities, learners are required to apply these skills to find out about words. Ensure that the glossary in *Cambridge Primary Science Stage 6 Learner's Book* is available and a hard copy or online dictionary is accessible. Check that learners remember how to use both and give support where required. Remind learners that they are using skills from another area of the curriculum in science.

Further activities

Challenge the class or individuals to find words beginning with 'macro', the prefix which has an opposite meaning to 'micro'. For example: macroscopic, macrofossil, macrobiotic, macropod.

ICT links

Give learners access to an online dictionary.

Revision quiz

Revision quiz tips

Prior to the quiz, discuss the importance of reading the questions carefully. Explain that sometimes the question gives clues about how to answer, for example, the vocabulary used could suggest the words to be used in the answer.

Answers

1 a

2 b

3 **a** True
  b False

4 **a** Circulatory system – Moves blood around the body.
  b Diaphragm – A muscle that moves up and down as you breathe.
  c Microbe – A tiny life form that can only be seen with a microscope (for example, bacteria).

5 To pass or spread an illness (disease) from one person to another.

6 **a** sneezing; coughing; not washing hands
  b the immune system fights against the cold; produces slimy mucus in nose and throat to capture microbes; makes us sneeze to blow the microbes out of the body

7 For example, whale and human heart pumps blood around its body, pumps oxygen, both have four chambers.

8 micro/microscope

9 **a** What is the process called when oxygen and carbon dioxide move between blood and the lungs?
  b What is the system called where the lungs and heart and other parts of your body are involved in breathing, and exchanging oxygen and carbon dioxide?
  c What is another name for when you inhale and exhale?
  d What is it called where the lungs and heart and other parts of your body are involved in breathing, and exchanging oxygen and carbon dioxide?

10 **a** What reasons do people give for not washing their hands regularly?
  b Nowhere to wash hands – it had the highest percentage – 36.2%.
  c Explain what happens if they do not wash their hands. Explain how easily disease is transmitted. Tell them ways that they can keep hands clean, for example, use hand sanitiser.
  d Display posters in washrooms, show adverts on TV, remind friends and family.
  e Washing with soap and water gets rid of most of the bacteria on hands and helps to stop viruses being spread through shaking hands, touching objects, etc.

11 The circulatory system moves blood around the body. The respiratory system exchanges oxygen and carbon dioxide in the body.

12 Both systems are involved with/have to do with exchanging oxygen and carbon dioxide in the body. (Accept moving oxygen around the body.)

Unit 2 Human reproduction

Human reproduction

Study Guide pages 24–25

Learner's Book pages 27–28

Objective

Learners will revise:

- the parts of the human reproductive system.

Science background information

Learning about the reproductive system and puberty will of course take place within the framework set by the school. This topic requires a thoughtful approach by the teacher, recognising that some individuals may know more than others and some learners might find the topic embarrassing. Ensure that learners understand that they should respect each other in their discussions and work.

The objectives provide two clear intentions, firstly that learners should be able to name the parts of the human reproductive system. The expectation should therefore be that learners can locate these on a diagram, name and be able to spell them.

The second objective requires that learners can describe the physical changes that take place during puberty. This requires a more nuanced approach since some learners in your class might already have begun to experience changes which some in the class might find awkward to discuss.

Throughout the Study Guide there is a focus on ways that learners learn, they should be encouraged to think about their thinking (metacognition), so do ask them to explain how they are learning words and information, which approaches work for them and why.

Revision approach background information

Key word cards are used to begin this topic to help prompt learners to remember key scientific vocabulary. Discuss with learners how they can divide the list of words up into those they know, are unsure about and do not know. You could agree that they only focus on those in their unsure and do not know lists. Encourage them to reflect on approaches that they find useful to help them learn words, for example, look, say, cover, say/spell. Explain that there is no correct way, it is up to individuals to find the best approach for them personally, for example, one learner might repeat how to spell a word 10 times, other learners might split the word up or write parts of it in different colours. Using cards means that if you have some learners who find the topic awkward, they can work individually.

Starter activity suggestions

Ask learners to write each key word on their cards and sort them into *Yes*, *No* and *Unsure* piles. They could use a dictionary or the glossary in the *Cambridge Science Stage 6 Learner's Book* to check the words that they think they know. If they are incorrect, they should move the card to another pile. It would be useful for learners to take a photograph of the words in each pile, so that later they can use it to reflect on their progress.

Activity notes and answers

Page 24

Activity 1: Answers

1 Check that learners have spelt each word correctly, whether they know the word or not.

2 Learners could use an individual whiteboard to draft and redraft their definition, or use two pieces of paper, their draft definition, and their final definition. Discuss what changes they made and why they think that their final definition is better.

3 This could be carried out individually, in pairs or even as a home school activity. Do make sure that you and learners are confident that they know and can define each word.

Page 25
Activity 2: Answers

2 Learners re-sort their word cards and create three new piles of *Yes*, *No* and *Unsure*. Ask them to compare their new pile with the photograph of the words in their original piles, then ask them to reflect on the progress they have made. Are there any words that they are finding very hard to learn?

3 The approaches will depend on each learner's personal learning style, but could include, for example:
   - Make two sets of cards: one colour are the words and the other colour the definition and play the game of pairs. All cards are placed face down, a word card is turned over then a definition card – if they match the player has a pair. If not, the cards are placed face down again in the same position and the other player has a turn. This continues until all pairs are complete.
   - Take the *Unsure* set of cards home to learn.
   - Use words in a sentence to indicate that they know what they mean.
   - Use sticky tack to put their cards into their book and move them into the *Yes* column as each word is learnt.

   The sentences will depend on the individual learner and the words that they choose.

Cross curricular

Discuss strategies used in Literacy to learn words and ask learners to think about which ones they find the most useful and apply them to learning words on the cards.

This activity helps to consolidate using dictionary and glossary skills.

Further activities

Place each word or definition on an A4 sheet of paper, then stick each sheet at a different point around the classroom. Learners then sticky tack their word or definition around the correct answer. Make sure that they put their initials in the corner of each one so that they can retrieve their own card.

ICT links

Learners use spell check on their computer to check spellings and dictionary to check definitions.

Reproduction and puberty

Study Guide pages 25–28

Learner's Book pages 27–28

Objectives

Learners will revise:
- the parts of the human reproductive system
- physical changes that take place during puberty in humans.

Science background information

The key learning here is that animals, including humans, need a male and a female to reproduce (have offspring). Animals can be divided into two groups, some animals grow their young inside their bodies and give birth to live offspring, humans do this. A second group lay eggs, from which their young hatch. In science, the use of prefixes and suffixes helps to define words, this is the case in animal reproduction:

The prefix **vivi** means living or alive.

The prefix **ovi** means egg.

The suffix **parous** means having produced offspring.

So **viviparous** means animals that give birth to live young, for example, humans (and most other mammals) while **oviparous** are those animals that lay eggs, for example reptiles and birds.

Revision approach background information

Learners will know from Literacy and their work in science about the use of prefixes and suffixes. The use of prefixes and suffixes can help in revision where learners know what a prefix means, for example, *micro* (small), *therm* (heat), this will help them to use their memory to work out what certain words mean when you read or hear them. Prefixes and suffixes were revised in Unit 1.

Starter activity suggestions

Remind learners about prefixes and suffixes, you could provide a list from other topics. For example:

micro – small   scope – instrument for viewing

therm – heat

Remind learners that they are also used in science where mathematics is used, for example in measurement, kilo – thousand, centi – hundred. Ask learners to find out what the prefixes *deci*, *centi*, and *milli*, mean (respectively they mean, one-tenth, one-hundredth, and one-thousandth).

Activity notes and answers

Page 26
Activity 1: Answers

1 **a** oviparous
  b viviparous

2 For example, any bird (lays eggs), reptiles such as snakes, crocodiles, lizards (lay eggs). Ask learners to explain why they have chosen these animals. Accept egg-laying (*oviparous*) mammals such as duck-billed platypus and spiny ant eater (echidna).

3 For example, humans, cats, elephants, horses, giraffes because they give birth to live young. Be aware that there are often exceptions to the rule, for example, some snakes, such as the Garter snake, are viviparous, so are some lizards and a few insects such as aphids.

Activity 2: Answers

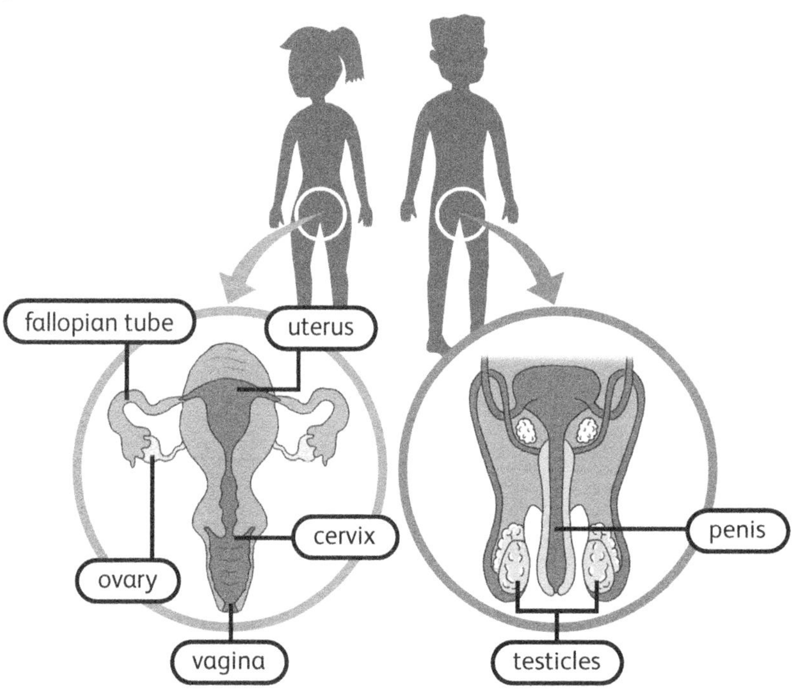

Page 27

Activity 3: Answers

1 For example: Gestation period – The time it takes for a foetus to develop, starting from fertilisation and ending at birth.

2 For example: Inherit – Process by which the information in cells is passed from parents to offspring.

Activity 4: Answers

2 a Correct

b Incorrect

c Incorrect

3 For example,
Our conclusion is that the bigger the animal, the longer the gestation period, for example the Orca has a gestation period of 510 days and has a weight of 136 000 kg, elephants have a gestation period of 645 days and a weight of 6 000 kg. However, a mouse has a much shorter gestation period of 19 days and only weighs 0.020 g, a cat has a gestation period of 60 days and weighs 4 kg.

Page 28

Activity 5: Answers

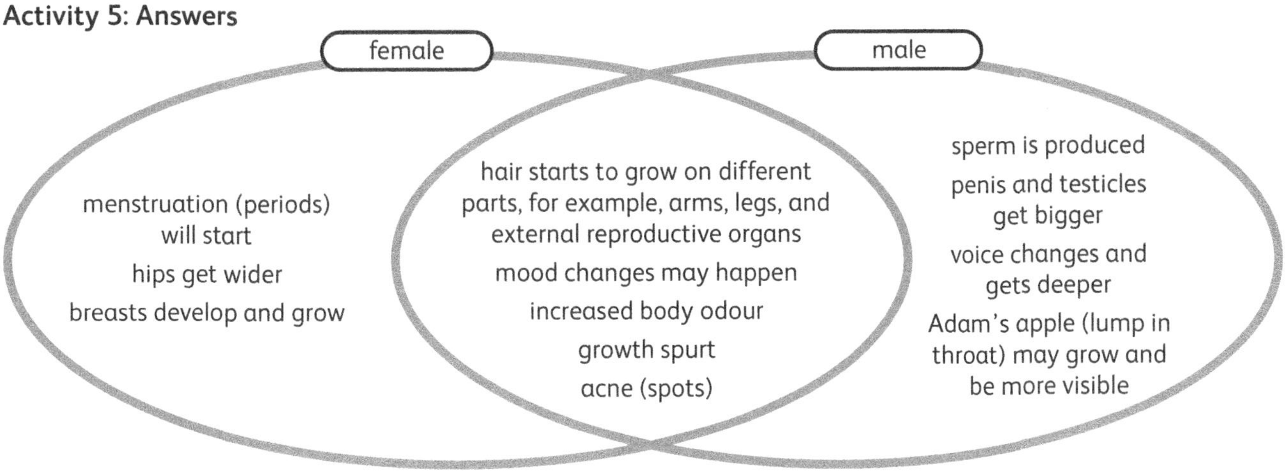

Activity 6: Answers

Accept appropriate responses from learners, for example:

Definition The stage of the human life cycle during which adolescents reach sexual maturity and become capable of reproduction.	**Characteristics/features/facts** female and male bodies change
Examples Females – menstruation (periods) will start; hips get wider; breasts develop and grow Males – hair starts to grow on different parts (arms, legs, and external reproductive organs); mood changes may happen; increased body odour	**Non-examples** they like to meet friends

Puberty

Cross curricular

Learners should have come across different kinds of graphic organisers in both Literacy and other subjects, for example, history and geography. Discuss the idea that a Frayer map is a type of graphic organiser, it helps to organise information and prompt learners to remember different aspects of a topic that they have learnt and organise where the information should go. Ask them to reflect on how useful it was in helping them to remember information about puberty, especially parts that they remember now that they did not before.

Further activities

Challenge learners to find out the names of as many oviparous and viviparous animals as they can that live in Pakistan.

ICT links

Learners use the internet to research oviparous and viviparous animals that live in their country.

Revision quiz

Revision quiz tips

Ask learners what kind of questions they find difficult and why. Make a list of the question types, for example, explanations, multiple choice and discuss ways in which they can help themselves when they have to answer those question types.

Answers

1 b

2 c

3 **a** True

b False

4 **a** Ovary – Where the eggs form in a female's body.

b Sperm – Made in the male's testicles.

c Pregnancy – When the egg has been fertilised by the sperm.

5 Any of the following:

Female – menstruation (periods) will start, hips get wider, breasts develop and grow

Male – sperm is produced, penis and testicles get bigger, voice changes and gets deeper

6 Any of the following: hair starts to grow on different parts (arms, legs, and external reproductive organs), mood changes may happen, increased body odour, growth spurt, acne (spots).

7 **Puberty** is the time during **adolescence** when a boy or girl's body begins to change and develop as they become an adult. Puberty is the time when humans reach sexual maturity. This process usually takes place between the ages **10 and 14** for girls and the ages **12 and 16** for boys.

When the body is ready to begin puberty; it releases special **hormones** (chemicals that tell the body what to do), these hormones change different parts of the body depending on whether you are a boy or a girl, some of the changes are the **same** for both.

8 **a** A graph to show the gestation period of different animals

b Name of animal

c Days

d Elephant, polar bear, camel

e Gerbil

f Half-way between 400 and 500 days

Unit 3 Ecosystems

Food chains

Study Guide pages 31–33

Learner's Book page 32

Objectives

Learners will revise:

- food webs and how to identify food chains within them
- energy sources of a food chain and food web and how energy is transferred through food chains and food webs.

Science background information

Food chains are a way of communicating how plants and animals get their energy. All food chains start with energy from the Sun. This energy is captured by plants during the process of photosynthesis. When asking learners to draw food chains, most chains start with a producer. A producer is an organism that can make its own food, therefore most food chains start with a green plant. Green plants make their own food through the process of photosynthesis.

In food chains, consumers are living things that cannot produce their own food, for example cows and lions, so they eat plants and animals. Consumers that eat other animals are known as predators, and the animals they eat are prey. When drawing food chains, the arrow always goes from left to right →. Learners should be encouraged to read food chains as if it is telling a story of how energy is transferred from one living thing to another. Encourage learners to use the terms 'energy' and 'transfer' rather than, for example, the antelope eats the grass, and the lion eats the antelope. So, each time learners see the arrow they should say either out loud or in their head, 'Energy is transferred from the producer, for example, grass, to its consumer, for example, antelope', or 'Energy is transferred from the antelope to its consumer, for example, lion'.

It is useful to remember that although food chains are shown as linear, most animals eat more than one type of food and therefore can belong to more than one food chain. To show the complexity in a habitat, food webs show different food chains and how they are interrelated and connect. Different habitats may have different food chains and therefore different food webs.

Starter activity suggestions

Give each group a set of cards with the names of plants and animals (ensure there is a variety of invertebrates, birds, reptiles, amphibians and mammals) and arrow cards. Different groups could have animals and plants from different ecosystems, for example, desert, ocean, rainforest.

Ask each group to use the cards to create a food chain, challenge them to see which group can make the longest chain. Once they have created their food chain, groups pair up and read each other's food chain. Then each group mixes up their cards and swaps them with cards from another group and the activity is repeated.

Activity notes and answers

Page 32

Activity 1: Answers

A Learners who understand energy transfer will use the following terms. For example:

Energy is transferred from the plant to the antelope, then energy is transferred to the lion.

Learners who are not confident might offer a different response, for example:

The plant is eaten by the antelope, the antelope is eaten by the lion.

This will allow you to assess the level of learners' understanding of energy transfer in a food chain.

Producers – bush/shrub/plant

Consumers – antelope, lion

B Energy is transferred from the plant to the grasshopper, the energy is transferred from the grasshopper to the mouse, and energy is transferred from the mouse to the owl.

Producer: plant

Consumers: grasshopper, mouse, owl

Page 32

Activity 2: Answers

Learners could add any insect (for example, grasshopper) that a bird would eat.

plant → [insect] → bird → snake → owl

Pages 32–33
Activity 3

Check that the food chain in learners' mini concertina books is correct.

Cross curricular

Learners could draw on their knowledge from geography of specific habitats, for example, coastal areas, lagoons, mangrove, swamps, sandy areas, grasslands, freshwater lakes, wetlands, mountainous ranges, valleys, snow covered mountains.

Further activities

Take learners into the locality to look for evidence of food chains, they could draw food chains onto the school playground using chalk, leaving them for other groups to discuss and comment on. Encourage them to take photographs of the habitats and evidence of food chains, for example, leaf damage, berries, feathers, spider webs, sightings of insects, birds.

ICT links

Learners could use the internet to research habitats in their own country or locality. They could take photographs of animals and plants in their own school grounds or home to use in their concertina books. Some learners could use computer graphic programmes to create their concertina books.

Food webs

Study Guide pages 33–34

Learner's Book pages 33–34

Objectives

Learners will revise:

- food webs and how to identify food chains within them
- energy sources of a food chain and food web and how energy is transferred through food chains and food webs.

Science background information

A food chain shows a single path of energy flow, showing for example, energy flows from a plant to the insect that eats it, and then to the bird that eats the insect and then to the hawk that eats the bird. If learners have not yet fully understood the concept of energy flow, then the food chain will be described as the plant is eaten by the insect that is eaten by the bird that is eaten by the hawk.

A food web shows the different paths of energy flow in a habitat/ecosystem. Like food chains the food web will begin with a plant and end with a carnivore.

Some learners will find it useful to think of food chains and food webs as telling the story of energy flow, a food chain is simpler because it only tells one story whereas a food web is made up of many interlinking food chains, so the story is more complex.

Starter activity suggestions

Give learners a piece of thick cardboard or corkboard, pictures of plants and animals, board pins and elastic bands. Using these resources, learners can make both food chains and food webs. Ask learners to explain their food chains and food webs to find out where they are in their learning.

Activity notes and answers

Page 33

Activity 1: Answers

There are at least three food chains within the ocean food web.

For example:

Sun → phytoplankton → krill → fish → leopard seal → whale

Sun → phytoplankton → zooplankton → squid → elephant seal → whale

Sun → phytoplankton → krill → fish → penguin → whale

There are at least six food chains within the desert food web.

For example:

Sun → desert plant → insect → scorpion

Sun → desert plant → lizard → desert fox → hawk

Sun → desert plant → insects → lizard → hawk

Sun → desert plant → insects → lizard → desert fox

Sun → desert plant → kangaroo rat → snake → hawk

Sun → desert plant → kangaroo rat → hawk

Page 34

Activity 2

Check that learners choose organisms that they do not know or know very little about. Discuss what kind of information should go on their fact cards and how it will be arranged (including a picture of the organism).

Revision approach background information

Activity 3

Remind learners that in Unit 2 they made key word cards to help them revise and remember key scientific words.

Explain that they are going to make a set of key word cards to help them revise and embed scientific words in their long-term memory,

Check that learners have correctly defined the key words using the glossary at the back of the *Cambridge Science Stage 6 Learner's Book* or a dictionary.

For example:

food chain – a diagram that shows (in a row from left to right) how organisms in a particular habitat feed on one another. Arrows are used in a food chain to indicate the direction that energy flows through it.

food web – a diagram made up of several food chains linked together that shows how living things in a habitat rely on one another for food.

toxic – poisonous

producer – make their own food, for example, green plants that make their own food

consumer – living things that eat producers and animals

energy flow – in a food chain or food web this is where the energy from the Sun moves to in an ecosystem/habitat as animals feed.

Revision approach background information

Model answer (Worked answer)

A model answer is an answer that has already been worked by someone else and is given to learners so that they can apply their understanding, reflect on how good the answer is and consider strengths and weaknesses. Learners engaging with this type of activity are revising their science because they must consider which aspects are good in the answer and where improvements can be made. Rewriting an answer after critiquing the original should provide learners with an insight into what a good answer looks like. If it is the first time that a group has carried out this kind of task, work through it as a class, explaining the reasoning for this kind of activity and how it can help learners, particularly when they must write extended answers in science.

Discuss with learners what they think should be the criteria for a model answer against which they can review the answer in the text. For example:

- Use scientific vocabulary.
- Compare similarities and differences between a food chain and food web.
- Explain key language such as consumer and producer.
- Use correct punctuation, grammar and use different connectives.

Page 34
Activity 4
Example answers:

2 Looking at this explanation I really like ...

Use of the words 'difference' and 'dependent', the answer is correct, they could have used the phrase 'energy flow' and more scientific vocabulary, for example, producer, consumer. Use of the words 'because' and 'whereas'.

This is what I would have written to improve the learner's answer ...

The difference between a food chain and a food web is that a food chain is a diagram that shows how organisms in a particular habitat feed on one another. Arrows are used in a food chain to show the direction that energy flows through it, whereas a food web shows lots of food chains and how they are linked together. Food chains and food webs are the same because they show producers (organisms that make their own food) and consumers that eat plants and animals for their food.

Cross curricular

The model answer activity is a form of peer assessment where learners comment on someone else's work, indicating what they liked/thought was good about it and what they could do to improve it. To do this, learners need to think about what the success criteria for the answer would be against which they can review the work. This is a strategy taught in Literacy, so spend some time discussing with learners that they are using an approach from another area of the curriculum, and that they already know how to think about success criteria and reflect on someone else's work.

Further activities

Challenge learners to continue their revision of food chains and food webs at home asking them to either create a food web relating to, for example, their garden, a park, or other local habitat (beach or urban landscape) and share it with others in their class. Or they could use the internet to research a food web in an area that they have not visited, for example, mountainous area or mangrove.

ICT links

Create a food chain or food web using a graphics programme to share with others.

Science in context – The Minamata Story

Study Guide pages 35–36

Learner's Book pages 36–40

Objectives

Learners will revise:

- the idea that some substances can be toxic (poisonous) and harm living things
- the idea that toxic substances can move through a food chain and food web affecting plants and animals
- food webs and how to identify food chains within them
- energy sources of a food chain and food web and how energy is transferred through food chains and food webs.

Science background information

The following activities focus on the story of Minamata, a fishing town in Japan where in the 1950s more than 1 700 people died and thousands were affected by heavy metal poisoning as a result of eating fish and shellfish from Minamata Bay. This was the result of bioaccumulation – 'bio' meaning life and 'accumulation' literally meaning something that has gathered over time, in this case a gradual increase in the concentration of mercury in the food chain. The source of the mercury was the Chisso plastics factory in Minamata which discharged methylmercury into Minamata Bay. Organisms, such as plankton were eaten by small animals and zooplankton, which in turn were eaten by small fish, then larger fish and then birds and mammals which includes humans. Learners will recognise the food chain, however what might not be clear to them is that larger fish, to stay alive, need to eat large quantities of smaller fish. This means that those animals ingest many fish that are contaminated with methylmercury and so the concentration of the poison is high in their bodies. When people living in Minamata ate fish caught in Minamata Bay, the fish they ate had high levels of methylmercury.

When considering the food chain, the greatest accumulation of the poison was in small fish and larger marine animals.

The example of Minamata shows science in a real-world context, and how toxins can move through a food chain, one which also involves humans, with devastating effects. The key idea here is not only that toxins can move through a food chain, but as they do the toxins accumulate to such a level that they are a threat to human life.

Revision approach background information

Rich picture poster

A rich picture is a way of showing an idea, information, a process or for example, a story, such as Minamata Bay, by using pictures, diagrams, individual words, phrases, and colour coding. Using a rich picture can sometimes be an easier and more interesting way for learners to show what they know than, for example, writing sentences or paragraphs, especially for those who are visual learners.

In the context of the story of Minamata Bay, the key issue is to ensure that learners understand that their rich picture should not only tell the facts of the story, for example, Chisso plastics factory, what the poison was, how many people died or were affected, but also the science behind the disaster. Challenge learners to include an explanation of the concept of bioaccumulation in a marine food chain.

Starter activity suggestions

Give learners time to read and discuss the story of the Minamata disaster, ask them what they think about the story, the effect on the people living there, how the poison got to humans, and what the key science ideas behind the story are.

Remind learners or explain to them the idea of creating a rich picture poster and discuss what the criteria for a 'good' poster would be, for example:

- Description of what happened.
- Explanation of the effect on people living there.
- Explanation of how the poison moved through the food chain and poisoned humans, using food chains/food web.
- What the effect of the poison was.
- What the term 'bioaccumulation' means.

Activity notes and answers

Page 36

Activity 1: Answers

Each rich picture will be an individual piece of work, so each one will be different. Use this as an opportunity for peer assessment where learners share their rich pictures and comment on what they like and what could be improved. Give learners sticky notes to write their comments which can then be placed on or around individual posters. Remind learners that they should keep referring to the agreed success criteria when they review each rich picture.

1 Methylmercury
2 The methylmercury in the sea was eaten by zooplankton and small animals, which were eaten by larger fish. Humans caught fish to eat and so were eating fish that was poisoned and many people became ill.
3 The top predators to stay alive had to eat lots of smaller fish, so they ate lots of fish that was poisoned. The poison stayed in their bodies and kept accumulating, so they had more and more poison.
4 Stop polluting seas and rivers, lobby government to make sure that industry does not pollute water.

Further activities

Show learners a video or transcript of a first-person account of the effects of the Minamata disaster, for example, a video from the internet. Ask learners to discuss what the impact was on the person's life and how they would feel if the same had happened to them. They could write a newspaper article about the incident, or a diary entry by a scientist.

Cross curricular

Ask learners to write and perform a short play about the Minamata disaster, they could show it to other classes and explain how environmental pollution and bioaccumulation can affect the food chain and humans.

ICT links

Learners could do online research to find out additional information about Minamata, create a newspaper article using a suitable programme or watch a video clip of a first-person account of the Minamata story.

Revision quiz

Revision quiz tips

Discuss with learners the idea that in some topics there are only a few key ideas that they need to learn, in this topic ideas related to:

- Energy flow in a food chain
- Food webs
- Bioaccumulation of toxins

Also emphasise that in this topic there are fewer key words to learn than in some other areas of science learning. The key words are:

food chain | food web | producer | consumer | toxin | bioaccumulation

Helping learners to recognise that some topics might be 'easier' than others to revise can help some feel less daunted by a quiz on the topic.

Answers

1 b

2 c

3 The food chain should be: Sun → grass → rabbit → eagle

4 b

5 Both show the flow of energy from producers to consumers. Food chains show the flow of energy from producers to consumers and food webs do the same by showing different food chains in the same habitat.

6 A food chain is very simple, it shows only one path of the flow of energy. A food web is very complex, showing many food chains and how they are linked.

7 Mercury in the water is taken in by small organisms such as phytoplankton and zooplankton, then eaten by small insects in the water. The small insects are then eaten by small fish, which are eaten by bigger fish and if those fish are eaten by humans, people could be poisoned. The bigger fish eat lots of smaller fish and so have more poison in them, the poison accumulates in the food chain, with more and more poison in bigger fish.

8 a Accept reasonable responses. For example:
plant → sea snail → crab → squid → tuna → shark
zooplankton → small fish → marlin
zooplankton → krill → small fish → shark

b marlin, shark, tuna

c For example, crab, lobster, water snail, shark, tuna, squid

9 Check learners' food web. Make sure that the food web includes humans.

10 In Minamata, the tuna had 10 million times the amount of methylmercury because of bioaccumulation, this is because the small fish that tuna eat are poisoned with the methylmercury.

11 The top predators are most at risk because to stay alive they must eat a lot of other animals which might have poison in them. The small animals have a little poison in them but because top predators eat so many of them, they accumulate the poison from all the fish they eat in their bodies.

Unit 4 Reversible and irreversible changes

Reversible and irreversible changes

Study Guide pages 38–39

Learner's Book pages 42–43

Revision hexagons template

Objective

Learners will revise:

- what they already know about materials.

Science background information

In science, it is important that learners not only learn information and facts but also wider concepts and that they are able to make links across a topic. The hexagon approach explained below is useful because it challenges learners to provide evidence to support their reasoning by visually connecting a series of ideas written on paper or digital hexagons around a theme.

This visual approach supports learners in not only communicating individual ideas but also articulating what connections they can make across ideas and therefore the depth of their learning. The range and detail of the connections they make enables the teacher to assess their learning. The use of hexagons supports learners in transferring learning into long-term memory where ideas are embedded and connections made between new and existing learning, which in turn helps them to make sense of their ideas.

Revision approach background information

Revision hexagons are hexagon shapes that tesselate (link together), the ideas and facts that learners write in them must also link to what is written in the surrounding hexagons.

This revision approach helps learners to:

- think about what they know
- recall facts and ideas
- make links between learning.

As learners write their ideas in each hexagon, this shows how they are linking ideas together and where they can make connections, providing the teacher with a useful formative assessment point.
Do let learners know that they can return to their hexagons at any point during this unit and add to them as they remember more during revision or extend their learning with new information. Using different colour pens is helpful, one colour for the hexagons completed in this first activity, and another for additions during and at the end of the topic to track learning.

Starter activity suggestions

For learners that have not used this approach before or need reminding, a trial run is very useful. Allow learners to work in pairs and provide them with copies of the revision hexagons template. Use a science topic that learners are very familiar with, it could be the previous science topic, for example food chains. Tell them to put the word 'food chains' in the centre hexagon and then write into an adjoining hexagon something that they know about food chains, they continue to do this but every time they write in a hexagon the information must link with all the adjacent hexagons. As they work, go around and check that learners understand how this revision approach works.

Activity notes and answers

Page 39

Activity 1

Accept all answers as this is a revision activity and will indicate how confident individuals are in what they know about the topic 'materials'. Some learners might remember straightforward ideas such as names of materials and their properties, for example, metal, wood, flexible, transparent. Others might be able to recall ideas relating to reversible and irreversible changes, dissolving, etc. This will provide useful formative assessment information and indicate whether some learners will require additional input.

Activity 2

This is an opportunity for learners to share what they have written and then question each other if they think a hexagon contains information that is not correct. They then talk through what they have written and change it. Self and peer assessment is a powerful tool in learning and in revision of personal knowledge. Giving learners time at the beginning to think through existing ideas is important.

Cross curricular

Using the revision hexagons is also useful in other areas of the curriculum. The more frequently they are used across subjects, the more confident learners will become in understanding how to use them and how they can support personal revision.

Further activities

Create a large set of hexagons for display and as a class use these throughout the topic to recall information and link key ideas.

ICT links

Teach learners how to make their own pages of hexagons so that they can use this approach themselves for revision across science and other areas of the curriculum.

Materials

Study Guide pages 39–41

Learner's Book page 42

Objective

Learners will revise:

- everything has matter, including gases.

Science background information

Everything is made up of matter, matter is anything that has volume (takes up space) and has mass. Having mass is a property of a material.

While many learners will recognise that water and solids have mass, many are likely to be less convinced that a gas has mass because it is invisible to the eye. One way to address this misconception and help learners understand and 'believe' that gas has mass is for them to carry out a simple activity using sparkling water. Gas has mass (weight) and in sparkling water the bubbles are bubbles of the gas carbon dioxide, so a bottle of sparkling water will weigh more than the same volume bottle of still water.

Learners are asked to weigh a container of sparkling water and leave it to become flat, which occurs when the sparkling water is left out and the carbon dioxide escapes into the air. The difference in results will give learners evidence that gas has weight.

You might think about introducing learners to the scientist Joseph Priestly 1733–1804 who discovered oxygen and the first fizzy drink, soda water (sparkling water) which is water with carbon dioxide also known as carbonated water.

Starter activity suggestions

To elicit prior understanding, ask learners to think about three statements from three different learners, you could give them names. Tell them that each learner has shared their ideas about whether gases have weight and ask them to vote on which statement they agree with, reminding them that they only have one vote. Collate the results, and you could create a bar graph showing the voting.

Learner A I think that a gas must have weight because it is matter so it must weigh something.

Learner B I think that gases are lighter than air, this means that they do not weigh anything.

Learner C I think that because gas is invisible it cannot weigh anything.

Then explain to the class that they are going to carry out an activity to find out which of the statements is scientifically correct and that when they have completed the activity the class will vote again.

Activity notes and answers

Page 40
Activity 1: Answers

1 a Sparkling water
  b The sparkling water has water and carbon dioxide (gas) so will be heavier than the still water which just has water and no carbon dioxide.
  c All matter has mass even gas, so the bottle with 500 ml of water will have extra mass because of the carbon dioxide.

2 For example: Pour sparkling water into a glass, weigh it and record the results. Swirl the sparkling water around in the glass or leave it for a while until the water goes flat. Then weigh it again and compare with the first results. The difference between the two weights will tell you how much carbon dioxide was in the drink.

Activity 2: Answers

A Condensation	**4** The process in which a material changes state from gas to liquid.
B Evaporation	**1** The process that happens at the surface of a liquid when it changes from a liquid to a gas.
C Freezing	**2** The process in which a material changes from a liquid to a solid.
D Melting	**3** The process in which a material changes state from solid to liquid.

Page 41
Activity 3

1 Check that learners have listed the key scientific words linked to materials and used them in their revision hexagons.

2 Check that learners have reviewed the ideas in the **Do you remember?** panel and where appropriate added or amended their hexagons.

Cross curricular

Hexagons are shapes that tessellate. Challenge learners to find different shapes that tessellate and to explain why hexagons were used instead of other shapes that tessellate, for example, hexagons give more opportunities to make links. Ask learners to find things that tessellate in nature and in the built world and get them to use one of the tessellations they find instead of the hexagon template, for example, a brick wall, or a tortoise shell.

Further activities

Go back to the statements from Learners A to C in the Starter activity above and ask the class to vote again, stress that if they have changed their mind that is fine, it means that their learning has moved on. Compare the new results with the first set. How are the results different? Who has changed their mind? Why? What have they done and learnt that made them change their mind? Use this opportunity to identify which learners still hold misconceptions. What is their thinking? Do they need to repeat the activity to help them better understand that gas has mass and therefore weight?

ICT links

Challenge learners to use the internet to find four interesting facts about Joseph Priestly to share with the rest of the class or friends and family at home.

Changing states

Study Guide pages 41–45

Learner's Book pages 43–57

Venn diagram template

Objective

Learners will revise:

- boiling and evaporation are different processes.

Science background information

This section focuses on changes in state relating to evaporation, condensation and boiling. There are similarities between these processes and differences, focusing on these helps learners to make sense of each process and ensures that learners are aware of the bigger picture.

Evaporation is the change in state from liquid to gas, this happens at the surface of a liquid and at a range of temperatures. **Condensation** is the change in state from gas to a liquid, it is the reverse of evaporation. Both are physical processes and reversible.

Liquid changes state in the process of **boiling**, from liquid to a gas, while evaporation and condensation can take place over a range of temperatures, water boils at 100°C / 212°F (dependent on pressure).

Boiling point is the temperature at which the entire liquid changes into a gas. This happens when there is enough thermal energy for the particles in the liquid to break the bonds between them and to separate and become a gas.

A misconception held by some learners, is that water vapour is a liquid (it is a gas), another name for this gas is steam. Knowing that steam is a gas is also a misconception that learners have, as they think steam is a liquid.

Starter activity suggestions

For safety reasons, this activity must be carried out by an adult. It is also easier if this is carried out with a small group of learners, so that they can stand at a safe distance away to observe the kettle boiling. Alternatively show a video clip of a kettle boiling to the whole class.

The reason for showing learners a kettle boiling is to challenge common misconceptions about water in its different states.

As the water in the kettle boils, it changes to a gas throughout the entire liquid. This gas is called water vapour, and water vapour is an invisible gas. This invisible gas could also be called steam.

As the water vapour gas comes out of the kettle spout it cools and turns back into a liquid, and these are the water droplets which are seen forming in the air.

Encourage learners to observe just above the kettle spout, they should notice that there is a gap before they can see the visible water droplets. The water vapour gas is the invisible gap just above the spout, and the liquid water is the droplets seen forming in the air.

Revision approach background information

A Venn diagram is a visual approach using circles to show the relationship between things, for example, similarities and differences between two concepts, in this case evaporation and boiling. It helps learners to think about the key features of each of the processes, what is different about them and what is similar.

Activity notes and answers

Page 41

Activity 1: Answers

1 Check to find out the commonalities amongst learners, i.e. what most learners know and anything that is never or rarely listed (indicating that those areas of learning might not be embedded).

2 Use these responses as formative assessment and make time to go over those ideas with learners to ensure that they understand before moving on to the next set of activities.

3 Challenge learners to make sure that they learn to spell those words and know what they mean – give a date by which they need to be learnt.

4 Check learners' revision hexagons and discuss any additions which might not be scientifically correct or poorly phrased.

Page 42

Activity 2: Answers

1 A: Solid

B: Liquid

C: Gas

2 Solid: the particles are close together. Liquid: the particles are a little further apart. Gas: the particles are further apart and can move around.

3

Particles randomly arranged but close.	B
Particles have regular arrangement and are close.	A
Particles randomly arranged and far apart.	C
Particles move slightly (vibrate) but stay in the same position.	A

Page 43

Activity 3: Answers

Accept any of the following in the diagram. Discuss any elements that the learners did not include.

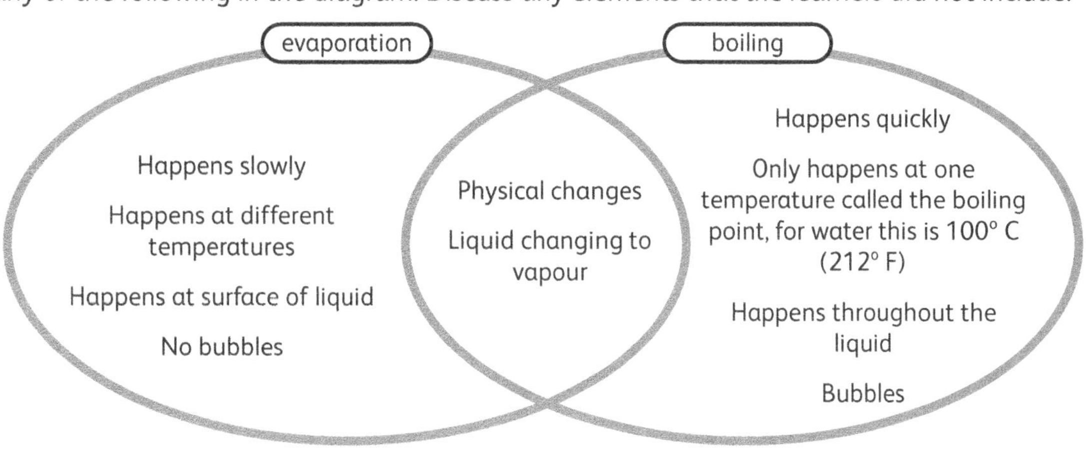

Activity 4

Examples of possible answers:

When learners have completed their fact files engage them in peer assessment where they comment on each other's work.

Process	What is it?	Reversible or irreversible	Explanation of your example of the process
Melting	When a substance is heated and changes from a solid to a liquid	Reversible	Chocolate melting because it has been heated
Evaporation	Water changes from a liquid to a gas on the surface of the liquid at different temperatures	Reversible	Puddle evaporating on a sunny day – the puddle disappears
Boiling	A liquid turns into a vapour when it is heated to its boiling point	Reversible	When a kettle of water is boiled
Freezing	When a liquid changes into a solid	Reversible	Making ice cubes – the liquid water turns into a solid – ice.

Pages 43–44
Activity 5: Answers

1 Scaffold learners in reading the graph. It does help if they understand that graphs tell a story of what happens, in this case containers of water and oil with different surface areas (you might have to revise the maths of surface area). Ask them to tell the two separate stories, firstly the story of the water and then the cooking oil. Get them to read the axis so that you know they understand what both axes represent, challenge learners to use data (numbers) from the graph.

   Ask learners to talk with a partner and describe what test they think the Stage 6 learners did to get the data for the graph. Encouraging learners to discuss the graph first before writing their answers ensures that learners articulate their thinking so that they and others can listen, decide if it matches the story shown by the graph and if not reconsider their ideas.

2 How does the surface area of the liquid affect how much water and cooking oil evaporates?

3 Using the same type of containers (volume and surface area) so that the surface area of the liquids are the same, same amount of liquid, same temperature/place, same amount of time.

4 Check their fair test map plan for the investigation.

5 Independent variable – surface area of the liquids

   Dependent variable – amount of liquid evaporated

   Control variables – using the same type of containers (volume and surface area) so that the surface area of the liquids are the same, same amount of liquid, same temperature/place, same amount of time.

Pages 44–45
Activity 6: Answers

1

Surface area	Volume of water at the end of 2 hours	Volume of cooking oil at the end of 2 hours
25 cm²	100	140
30 cm²	90	130
40 cm²	70	120
50 cm²	30	110

2 For example, if the question was, How does the surface area of the liquid affect how much water and cooking oil evaporates?

   They could answer: The greater the surface area the more liquid evaporates. More water evaporates than cooking oil. The cooking oil is a thicker liquid (more viscous than water), so it takes longer to evaporate, so less evaporates in the same amount of time.

3 Measure the liquid every 15 minutes, repeat the fair test.

4 The graph shows that the water and cooking oil evaporated because when the surface area was changed the volume of water changed.

Note *that learners might say that the graph shows the amount of water left, but there is no indication on the graph of how much water there was at the beginning of the test.*

5 Accept reasonable responses. For example: carbonated water (to see if bubbles in the water make a difference), hot and cold water (to see if already hot water evaporates quickly), comparing cooking oil with other viscous (thick) liquids such as golden syrup.

6 Check learners' tables. An example is included below.

Liquid	Prediction (Faster or slower than cooking oil)	Reason
1. Golden syrup	Slower	It is more viscous (thicker) than oil.

Cross curricular

Give the learners the responsibility of videoing the boiling water demonstration. Show the video to the class. Working in pairs, ask learners to create a commentary for the video clip. Discuss what the criteria for a good commentary would be, for example, use of correct science, linked to different parts of the video clip, key learning points. Get learners to put their commentary on a large sheet of paper, display them around the classroom and ask learners to leave sticky note comments, for example - one good comment, one suggestion for improvement and one comment linked to the criteria.

Further activities

Challenge learners to give examples of where they have seen evaporation, condensation and boiling take place and how they know that the examples they are suggesting are correct.

ICT links

Learners could create a picture collage of examples of condensation, evaporation and boiling and use a computer to draft and re-draft explanations.

Dissolving

Study Guide pages 45–47

Learner's Book pages 58–60

Graph paper

Objective

Learners will revise:

- temperature can affect how a solid dissolves; and we can use the particle model to describe how this happens.

Science background information

When some substances are placed in a liquid such as water they dissolve, to learners these substances such as salt and sugar look as though they have disappeared. These substances have not disappeared, they have been broken down into smaller parts and have mixed with the water which looks transparent.

Substances that dissolve in water are called **soluble** substances, sugar and salt are soluble. Sand is **insoluble**, it does not mix with the water, instead it settles at the bottom of the container. Where soluble substances such as salt or sugar dissolve into water, a **solution** is formed. Some substances such as flour do not dissolve in water and they do not sink to the bottom of the water, they create a cloudy mixture known as a **suspension**.

The substance that dissolves in a liquid is known as a **solute** (salt and sugar are solutes) and the **solvent** is the liquid (substance) in which the solute dissolves, water being the most common solvent.

The process of dissolving can be speeded up or slowed down. For example, dissolving sugar into tea can be speeded up by changing the temperature of the solvent (the tea), adding less sugar (the solute).

Learners may find it difficult to imagine what happens at particle level when a solute (sugar) dissolves into a solvent (water).

The simplest explanation of why substances such as sugar (solute) are dissolved in a liquid such as water (solvent) is because the solvent collides with the solute and breaks it down into smaller parts which can no longer be seen. At a more complex level of understanding, substances dissolve in water because the molecules of the liquid and solid are attracted to one another. When a solution reaches the point where it cannot dissolve any more solute (sugar), it is known as a saturated solution.

Dissolving is an area where a common misconception is sometimes held by both learners in the classroom and adults. Many people incorrectly say something is dissolving when it is melting and conversely say something is melting when it is dissolving. It is useful to discuss this with learners to ensure that they understand the difference between the two, a simple rule to teach them is that:

- For something to dissolve it needs a solvent. For example, sugar needs water to dissolve.
- For something to melt it needs heat. For example, chocolate needs to be heated to melt and cover a marshmallow.

Starter activity suggestions

The list below are the key words relating to dissolving. Give learners the words in a scrambled form and ask them to unscramble them and draw one or more diagrams on their mini whiteboards to illustrate each word. Engage learners in peer assessment, ask them to share their diagrams with someone else so that they can check each other's work.

dissolve | insoluble | soluble | solute

Activity notes and answers

Pages 45–46
Activity 1: Answers

1 When sugar dissolves it does not disappear, it is still in the water, it just cannot be seen.

2 The learner does not understand that when the sugar dissolves it is broken down into smaller particles by colliding with the water, the small particles of sugar mix throughout the water particles and the sugar is still in the water.

3 Some learners might suggest using clear beads or marbles, pouring 'clear beads' representing sugar into the 'marbles' representing water to show how the sugar particles mix with water particles.

Page 46
Activity 2: Answers

1

Substance	Soluble	Insoluble
Coffee	X	
Sugar	X	
Sand		X
Salt	X	
Pasta		X
Pebbles		X
Vitamin tablet	X	
Rice		X

2 Give learners the opportunity to carry this out, remember that just telling someone the answer does not encourage deep learning, when learners experience it for themselves, they are more likely to understand and remember, supporting deeper learning. If learners place items in the unsure column, remember that this is an appropriate response for them, and it can help the teacher to understand where they are in their learning.

Page 46

Activity 3: Answers

Accept reasonable answers. For example:

Prediction (including reason)	Resources	What to change	What to keep the same	What to measure
• The higher the temperature of the water, the faster the sugar will dissolve. The particles are moving quickly and banging into each other or moving away from each other.	• thermometer / digial thermometer • water • measuring jug • scales • salt or sugar • beakers	• Temperature	• Amount of water (solvent) • Amount of solute e.g. sugar	• Time to dissolve

1 Accept reasonable answers, for example:

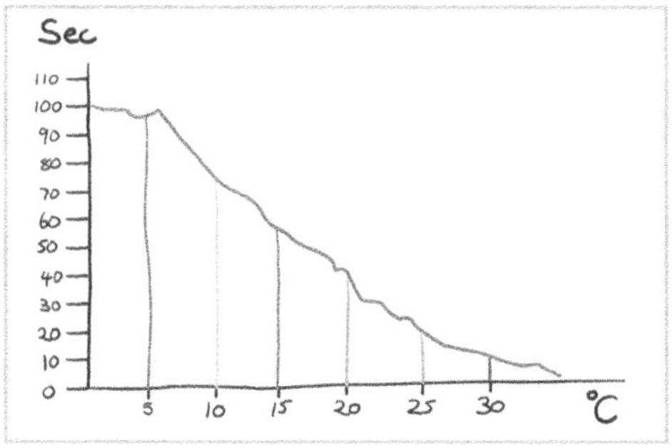

2 Their diagram should show how the particles move quickly and slightly further apart because they are being heated so dissolving occurs more quickly.

Cross curricular

This is a good opportunity to check that learners are applying maths skills correctly in their science. Ask them to list the type of maths that they need to think about when planning their fair tests. For example:

- Measurement (capacity, time)
- Repeat readings/calculations/average
- Line graphs – axes, line of best fit

Further activities

Challenge learners to suggest other questions that could be investigated using a fair test.

Encourage learners to use scientific language and measurement in their question, for example, How does the amount of solute affect the time it takes to dissolve?

ICT links

If digital thermometers, scales and stopwatches are available make sure learners use this equipment to ensure greater accuracy and finer measurements.

Reversible and irreversible changes

Study Guide pages 47–50

Learner's Book pages 61–63

Objectives

Learners will revise:

- physical changes are reversible (can be changed back to what they were before)
- chemical changes are irreversible (cannot be changed back to what they were before)
- in a chemical reaction substances called reactants act together to form new substances, these are called products.

Science background information

In this section the focus is on irreversible and reversible changes.

There are two key elements to the concept of irreversible change; the first is that when an irreversible change takes place the substance cannot be easily changed back to its original form, the second is that during an irreversible change a new material is formed, and these new materials can be useful. This of course is the opposite of reversible changes where the change can be undone or reversed easily.

Check that learners understand that where a reversible change occurs it might change how a material looks or feels, but it does not create a new material. Learners should appreciate that in a reversible change, for example, burning, there is a chemical reaction/irreversible change as the material reacts with oxygen to create the products ash and smoke, which are very different to the original material.

Examples of reversible changes	Examples of irreversible changes
Freezing water	Heating (making toast, cooking a raw egg)
Dissolving	Mixing (bicarbonate of soda and vinegar, baking a cake)
Evaporation	Burning (burning wood, paper, coal, oil)
Melting ice	Rust
Melting butter, chocolate or wax	Ripening fruit
Boiling water	
Stretching an elastic	
Inflating a balloon	

There are five key scientific words for learners to remember in terms of spelling, meaning and understanding examples, they are:

Reversible change: A change that can be undone or reversed.
Irreversible change: A change that cannot be undone, it cannot be changed back again, and new materials are always formed.

Chemical change (reaction): An irreversible change is called a chemical change (reaction).

Reactants: The materials that take part in and change during a chemical change are called reactants.

Products: In a chemical change one or more new materials are usually formed. These new materials are called products.

Starter activity suggestions

Discuss reversible and irreversible changes with learners to ensure that they know that reversible means they can get it back to its original form; irreversible means a new material is made and cannot get back to its original form (for example, bread – toast). Discuss arrows as a one-way process or reversible, two-way process. Use the word 'process' to describe reversible and irreversible changes.

Activity notes and answers

Page 47
Activity 1: Answers

1 → Irreversible: This arrow goes one way – forward, so it shows that it is not reversible.
←→ Reversible: There are two arrows – they go forwards and backwards which means the change is reversible.

2 Ask learners to share their hand signs with a partner. What do they think of each other's signs – are they easy to do, to remember, do they make the other person think of reversible and irreversible?

Pages 47–48
Activity 2: Answers

Process	→ or ←→	Describe two changes that take place.	Why is the change → or ←→ ?
Burning match	→	Smoke, flame, wood blackens, heat	The match has changed, it is black, smaller and cannot be changed back to before it was lit.
Making an ice lolly	←→	Gets colder, liquid solidifies	The ice can be melted turning from solid back to liquid.
Melting chocolate	←→	Gets warmer, chocolate changes from solid to liquid, becomes runny	The chocolate changes from solid to liquid and if cooled can be changed back to solid.
Rusty iron nail	→	Changes colour, metal changes and becomes flaky	Rusty nail cannot be changed back to what it was before it rusted.
Condensation on a mirror	←→	Mirror is not as reflective, water droplets form on mirror, droplets run down mirror	Condensation on mirror evaporates into the air, changing back to a gas.
Dissolving salt	←→	Salt appears to disappear into water, water tastes salty	Salt dissolves into water but if water is evaporated the salt will be left.
Baking a cake	→	Mixture goes from runny to firm, changes colour, smells different, gets bigger – rises, heats up, gets hot	Cannot get the solid cake's different products back to when it was a runny liquid.
Toasting bread	→	Becomes harder, changes colour, smells different, gets warmer	Cannot get the toast to go back to being soft bread and the colour it was before it was changed when heated.

Page 48
Activity 3: Answers

Irreversible change	Reactant	Product
Bread baking	flour, yeast, water	bread
Egg frying	raw egg, oil	cooked egg
Toasting bread	slice of bread	toast

Pages 48–49
Activity 4: Answers

1 Irreversible because when the bicarbonate of soda and vinegar were mixed there was a reaction and a new material, carbon dioxide gas/carbon dioxide bubbles were made. We also cannot get the bicarbonate of soda and vinegar back.

2 Yes, because a new material was made: carbon dioxide gas/carbon dioxide bubbles.

3 Bicarbonate of soda and vinegar are the reactants.

4 The product is carbon dioxide gas.

5 The glove was filled with or inflated by the carbon dioxide gas and the glove moved upwards because of the force of the gas that was produced.

6

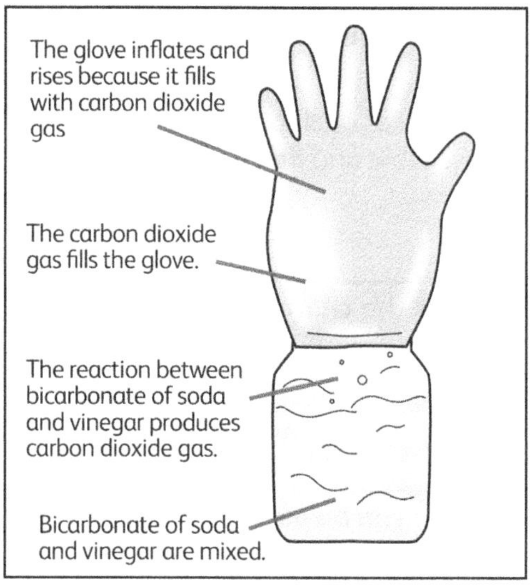

Page 49
Activity 5

Give learners time to think about which approach or approaches they will use to learn the key words. Ask them to self-assess and decide which words they already know and those which they still find difficult to say, spell and give examples of to someone they are working with, so that they can test their memory.

Page 50
Activity 6

Check the information that learners have added to their hexagons, ensure that the information they add in each new hexagon links to the hexagons around it. If you are unsure about any additions they have made, ask learners to explain what they have written.

Cross curricular

This activity provides an opportunity for learners to draft, edit and redraft explanations. Discuss with learners what the prefix ir- means (it means not) and how using it changes a word, for example, responsible when ir- is added becomes *irresponsible* which means not responsible. Link this back to the words reversible (can be changed back) and adding the prefix ir- to make the word irreversible, it changes reversible to not reversible.

Challenge learners to find six words which then change to the negative when ir- is added, for example, regular to irregular, rational to irrational.

Further activities

Challenge learners to ask questions that could be investigated by changing the variables when bicarbonate of soda is mixed with vinegar, encouraging learners to use scientific language, for example:

- How does changing the amount of vinegar/bicarbonate of soda affect the amount of carbon dioxide produced?
- How does changing the type of vinegar affect the reaction?

ICT links

Learners could create a picture collage of photographs of reversible and irreversible changes.

Thermal conductivity

Study Guide pages 50–51

Venn diagram template

Objective

Learners will revise:

- some materials are thermal conductors, this is a property of the material.

Science background information

A **thermal conductor** is a material that allows thermal energy to pass easily through it, for example, metal. A **thermal insulator** is material that does not let thermal energy travel through it easily, for example wood, rubber and plastic. Thermal conductors and thermal insulators are opposites. Do insist that when learners talk about this property of material that they use the complete term, i.e. thermal insulator and thermal conductor. This is important since there are other kinds of conductors and insulators in science, such as electrical conductors and electrical insulators. Remind learners that materials are chosen for a reason, for example, material that is a thermal insulator would be chosen to make a pair of oven gloves, to ensure the heat from the hot object cannot transfer to the person holding the pan. This is to make sure that learners understand that the science they are learning has everyday applications.

It is useful to help learners link their science knowledge across different topics, for example in this context, discuss with learners the idea that those materials that are good thermal conductors are good electrical conductors as well. For example, copper is both an excellent thermal conductor and electrical conductor.

Starter activity suggestions

Discuss the word 'conductor' with learners to elicit their understanding of the word, separating everyday use (bus conductor, conductor of an orchestra) from its scientific meaning (thermal conductor or electrical conductor). Discuss what the opposite of a conductor is in science, finding out what learners understand by the word 'insulate' and 'insulator'. Ask learners to write their explanations of what an electrical conductor and an electrical insulator is, including examples of materials. Tell them to swap with either another pair or individual to peer assess each other's work, leaving a star (something they like about their explanations) and a wish (something that they think could be improved).

Remind learners that electrical conductivity is a property of a material and so is thermal conductivity.

Give learners suitable magazines or old shopping catalogues and ask them to cut out pictures of thermal conductors and insulators and place them in a Venn diagram, or give them a range of small swatches and pieces of materials to sort and glue onto the Venn diagram, for example, different fabrics, plastic, aluminium foil, wire, pieces of wood.

Activity notes and answers

Page 50

Activity 1: Answers

1. Thermal insulators are materials that do not let thermal energy pass through them easily unlike thermal conductors which do let thermal energy pass through them easily.
2. Thermal insulators: plastic, wood, polystyrene, fabric, rubber, ceramic, paper, wool.
   Thermal conductor: any metals.

Page 50
Activity 2: Answers

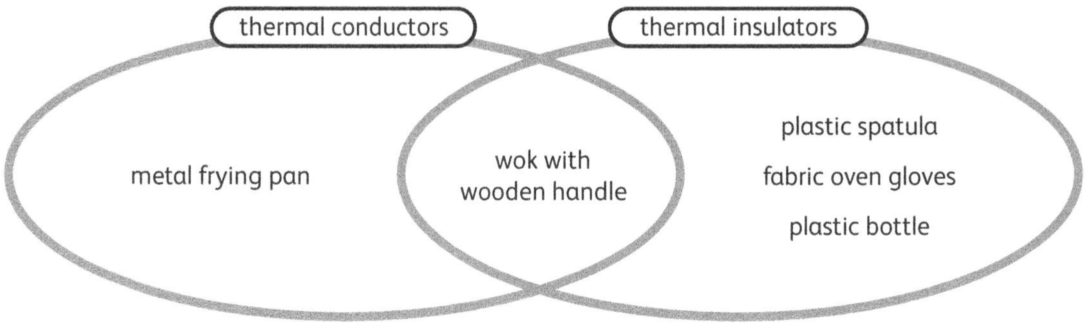

Page 51
Activity 3: Answers

1 Learners were trying to show that some materials are thermal insulators and some are thermal conductors. When placed in hot water the wooden and plastic spoons would not feel hot because wood and plastic are thermal insulators, but the metal spoon would feel hot because metal is a good thermal conductor.

2

Spoon material	Prediction	Reason
Metal	The spoon will feel hot to touch	Metal is a thermal conductor
Plastic	The spoon will not feel hot to touch	Plastic is a thermal insulator
Wood	The spoon will not feel hot to touch	Wood is a thermal insulator

Cross curricular

Discuss the idea that many of the skills learners learn in other subjects can be applied and further developed in science. For example, challenge learners to explain how they could apply their knowledge and skills in maths and ICT in science to testing different spoons for thermal conductivity. For example, maths could be used to measure the temperature of the water and each of the spoons, and a graph could be produced of the results, including a graph over time. ICT could be applied by using data loggers to log the temperature of each spoon, including over time.

Further activities

Challenge learners to design a lunchbox or bag to make sure that food keeps cool on a family picnic. Give learners the opportunity to research cool boxes and bags to find out what materials are used. They could create a design page with an annotated diagram of their cool box/bag and explain the choice of materials. Encourage learners to bring cool bags/boxes from home so that they can explore how they are made, and the materials used. If they make their designs, they could picnic in the school grounds to test how successful their cool boxes or bags are.

ICT links

Learners could research the inventor of the vacuum flask (James Dewar) and explain how it works applying their understanding of thermal insulators and thermal conductors. They could also carry out Activity 3 using data loggers to find out if their predictions match their original results.

Science in context – Global warming

Study Guide pages 51–52

Objective

Learners will revise:

- some materials are thermal conductors, this is a property of the material.

Science background information

This section focuses on Science in context, providing learners with access to information on how scientific ideas relating to thermal insulation are applied across the world in contexts which might apply in their own lives.

Thermal insulators are important in reducing thermal energy loss and gain in buildings across the world to reduce energy consumption, whether it is to reduce heating to keep a building warm or to reduce air conditioning to keep a building cool.

Thermal energy travels from hot to cold, so when buildings are heated, thermal energy will escape from any uninsulated area to the cooler temperature outside, this of course is equally important when keeping buildings cool. Thermal energy from outside will move from hot to cold (inside a building) and so air conditioning has to work harder to keep the building cool, using more electricity.

Across the globe, countries are focused on reducing energy use to heat and cool buildings, which include homes, where a third of thermal energy can be lost through walls, resulting in increased energy use. More effective insulation will result in less energy use and help to reduce carbon dioxide emissions which are a key contributor to global warming.

While fibre glass has been a common material used to insulate buildings; it can contain carcinogens (cancer causing material) and the process of making fibreglass is less green than using materials such as wool and cork. Denim is also becoming a popular material for insulation, not new denim material, but recycled denim fabric from clothes that have been thrown away and sent for recycling, which of course reduces the presence of this material in landfill.

Revision approach background information

Engaging learners in Science in context information and activities provides opportunities for them to apply their learning about thermal insulators to new and relevant world contexts. Being able to understand how their learning in science is relevant to the everyday world is important. Understanding the properties of different materials and their impact on the environment, develops learners' understanding of how this relates to global issues which challenge countries across the world.

Starter activity suggestions

Give learners a set of pictures, for example, a double glazed window and a polar bear. Tell them that both are good thermal insulators and ask them to work together in pairs or a group to explain why. For example:

Double glazing works by trapping air (usually the gas argon) between the two panes of glass, the gas acts as an insulator, the trapped air acts as a barrier between the inside and outside pieces of glass. The gas acts to reduce transfer of heat between the outside and inside of a building, allowing buildings to remain cool in summer and warm in winter.

Polar bears live in habitats that have extremely cold temperatures during winter months, as low as -30°C, so they need to be well insulated. Polar bears have a thick layer of blubber (fat) underneath their skin and a thick layer of fur (which is also waterproof).

Activity notes and answers

Page 52
Activity 1

Scaffold learners in their preparation for creating the slides. Give learners time to plan their four slides, so that they draft and redraft material including text, diagrams and pictures. Tell learners that before they begin, they need to create a list of key scientific words that they should use, for example, thermal insulators and thermal conductors, thermal energy, global warming, insulation, wool, denim, straw, heating, cooling, buildings.

Ask learners to list the most important ideas/information that they want to communicate/share on each slide before they begin and to think about the best way to share that information, for example, text, diagrams, photographs, data.

Page 52
Activity 2: Answers

This activity focuses on the use of prefixes, a common theme in the Study Guide. Where learners are unsure of the word/s encourage them to use a dictionary.

1 The prefix 'therm' means heat.

2 Thermographic scanners – a scanner used to detect temperature patterns and blood flow in the body.

Thermal imaging – a camera that uses the thermal energy given off by an object to produce an image of it. It is often used to locate humans, for example in areas hit by an earthquake where humans are buried in collapsed buildings.

Thermogram – a picture produced by a thermal imaging camera.

Thermographer – someone who understands and is qualified to use thermal imaging in their job.

3 The house needs insulating in areas where there are red, orange and yellow areas, for example, front wall and side wall because they are showing that thermal energy is being transferred from the inside to the outside.

4 The house already has insulation in the loft because it shows blue – cold. This means that thermal energy is not being transferred from the loft space through the roof to the outside.

Cross curricular

Climate change and pollution are key problems facing most countries. These are global issues which means that these are concerns that affect countries across the world. Challenge learners to think about how the basic idea of thermal insulation can be applied to their own homes either to keep them warm or cool. Widen this to buildings in their locality and the global impact on climate and the world if countries do not ensure that current and future buildings have effective thermal insulation.

Further activities

Challenge learners to think about their own homes. Are they insulated? If so, where and what kind of materials have been used? Ask them to think about which materials could be produced and used locally to insulate homes in their region to keep them cool or warm.

ICT links

Use the internet to show learners thermograms. Ask them to 'read' the thermogram and describe where cool, warm and hot areas are located.

Tips for success

This topic on materials covers a wide range of content from dissolving to changes in state, thermal insulators and thermal conductors. It would help learners if they not only referred to their hexagons but also created a memory map to organise different learning by grouping ideas under key areas such as dissolving, changes in state, thermal insulators and thermal conductors. Organising information in a different way to the hexagons can help learners to access different understanding in their memory and indicate where there are gaps. Check for gaps and misconceptions prior to the revision quiz and use discussion to help learners with their understanding.

Revision quiz

Answers

1 b

2 b

3 bread dough → bread

4 d

5 Boiling and evaporation are physical changes. In both, the liquid changes to vapour (gas).

6 The temperature of the water affects how quickly a substance dissolves because heating the water causes the particles/molecules to move and vibrate faster. This means that there are more collisions with the solute, resulting in it breaking up quicker.

7 Accept appropriate answers. For example: coffee, hot chocolate, lemonade, cola, orange squash.

8 c

9 Accept reasonable answers. For example: burning wood, rusting nail, frying an egg, baking bread or cakes.

10 Reactants – The materials that take part in and change during a chemical change.
Chemical change – An irreversible change.
Products – In a chemical change one or more new materials are usually formed.

11 c

Unit 5 Forces

Forces

Study Guide pages 54–56

Learner's Book pages 68–70

Objectives

Learners will revise:

- the use of force diagrams to show the type, size and direction of forces acting on an object
- different forces have different effects on an object at rest and in motion (moving).

Science background information

A force is a push or a pull. The existence of a force always requires an interaction between two objects. For example, a hand pushing on a box will exert a force on the box. This can affect the motion of the box, along with other forces acting on it, such as the force between the floor and the box. The two objects that interact will either be touching, these are known as **contact forces**, or interacting at a distance, these are known as **non-contact forces** (for example, magnetism).

Learners may find it conceptually difficult to understand that air and water can also exert forces due to their particles interacting with objects. It is therefore helpful to consistently name both objects that are interacting to cause the force to act. This is better than saying to learners that forces always act in pairs, as this is not always the case. However, by getting learners to always name the two interacting objects, they develop a better understanding of forces and how they affect the motion of an object.

Learners should already be familiar with a number of named forces: **normal force** (the force that a surface exerts to prevent solid objects passing through it), **applied force** (a force applied directly to an object by a person, animal or object), **upthrust** (the force that pushes up on an object in water or air), **gravitational pull** (gravity, the pulling force exerted by one mass on another; the greater the mass, the greater the gravitational pull between it and the mass it is interacting with), **friction** (a force that opposes the motion of an object), **air resistance** (a force that acts on an object as it moves through the air), **water resistance** (a force that acts on an object as it moves through the water) and **magnetic force** (a non-contact force with which a magnet pulls certain materials towards itself).

Forces have a size (magnitude), and act in a particular direction. Scientists use a force diagram to show the forces acting on an object. Force diagrams use arrows to represent forces. In a force diagram, the direction the arrow is pointing shows the direction the force is acting. The point at which the arrow is anchored also indicates how the two objects are acting. If the two objects are touching (contact forces), the arrow will be anchored at the point that the objects touch. If the forces are acting at a distance (non-contact forces), the arrow will be anchored at the centre of the object. The stronger the force, the longer the arrow.

Starter activity suggestions

To support learners, you could provide them with some examples of key word cards and explain the importance of reading out and sounding key vocabulary as well as adding images to their cards as this will help them to recall the word. A key point to this approach is that the learners create their own cards and are not given them pre-made. It is the process of making the cards that will help them remember and recall the ideas. Encourage learners to make additional cards as they encounter new key vocabulary throughout the unit. It is also useful, periodically, for learners to get out their cards and test themselves to see whether by just looking at the images they can recall the other information on the cards, or by looking at the words they can name the image. All of this will help consolidate the ideas in their long-term memory.

Revision approach background information

Key word cards are used again in this topic, however, this time the learners are asked to add images to the front of their cards. This additional image will help them recall and remember what they have put on the card. By using images in this way, the learners are dual coding the information they are creating. Dual coding provides two different representations of the information, both visual and verbal, when learners speak out the words. The use of images linked to the spoken words has been shown to help individuals recall and remember information they are learning about.

Activity notes and answers

Pages 54–55
Activity 1

Check that learners understand how to make their key word cards and note which words they have difficulty with. Check if this is common across the class, if so, discuss.

Page 55
Activity 2

The aim of this activity is to help learners remember the four rules related to best practice when drawing force arrows. Ask learners to focus for 20 seconds and then spot which rule is missing when they deal out three, this will help them recall and remember the four rules and solidify them into their long-term memory.

The four force arrow rules are that the arrows must:

- show the direction in which the force is acting. This is shown by the direction the arrow is pointing.
- show how strong each force is. The longer the arrow, the stronger the force.
- show where the force is acting from. The base of the arrow (the flat end) shows us where the force is acting from.
- be labelled with the name of the force, including the two objects that make up this force.

Activity 3: Answers

Marks should be given for:

- The support force arrow is pointing upwards, and the weight arrow is pointing downwards.
- Both arrows are the same size indicating that the forces are balanced/equal.
- The support force arrow is anchored on the table surface acting upwards on the cat as it is a contact force. The weight force arrow is anchored from the centre of the mass of the cat and acting from there as it is a non-contact force.
- The support force arrow is correctly labelled with the correctly named force 'support force' and it says which two objects cause it, the *table* acting on the *cat.* The weight force arrow is correctly labelled with the correctly named force 'weight' and it says which two objects cause it, the ***Earth*** acting on the ***cat.***

Page 56
Activity 4

Drawing force arrows to clearly identify the two objects and how they are interacting is difficult. However, supporting learners to accurately draw force arrows helps them develop the conceptual understanding of how forces affect objects. This activity is designed to check whether learners have fully grasped the key ideas and can apply them to incorrect situations.

Answers

1–2 The mistakes in each diagram are:

A The arrow pointing to the left in the water should be anchored at the front of the boat as this is where the water pushes back on the boat and produces the force of water resistance. It should be labelled '*push force of water on the boat*'. The arrow pointing to the right should be anchored on the sail as this is where the air pushes on the sail to create the forward push force, it should be labelled '*push force of wind on the sail*'.

B The arrow pointing up should be anchored where the air pushes up on the seed and be labelled '*push force of air on the seed*'. The arrow pointing down should be anchored in the centre of the 'fat seed' as this is where the gravitational pull of the Earth acts on the seed to create the weight force and be labelled '*weight force from the pull of the Earth on the seed*'.

C The arrow pointing to the left in the air should be anchored at the front of the motorbike as this is where the air pushes back on the motorbike and produces the force of air resistance. This arrow has been labelled correctly. The arrow pointing to the right should be anchored on the engine as this is where the engine pushes on the motorbike to create the forward push force.

Feedback comments the learners write should include reference to the four force arrow criteria and how the diagrams do not meet these and what they need to do in future examples to ensure they do not continue to make the same mistakes.

Cross curricular

Discuss with learners how when working in science it is important to use the correct language and ensure that diagrams are correct. This is particularly important in this topic, placing the arrow in a different position can change the meaning of the diagram. Ask learners to think about other subjects where it is important to use correct language and diagrams, for example, design technology, maths, geography.

Further activities

During PE lessons as learners carry out activities, for example, hitting a ball with a bat or kicking a ball, stop them and ask them to draw the force diagram for that activity on an individual whiteboard. Ask learners to peer assess. Do this 2 or 3 times during the lesson, applying their science to physical activity.

ICT links

Show learners different pictures from the internet, for example, person skiing, swimming, boat sailing, parachute open and falling. Ask learners to describe the forces.

Then ask them to choose four of their own pictures from the internet and use the draw facility to place force arrows and the name of different forces. Use peer assessment, asking them to check each others work.

Balanced and unbalanced forces and motion

Study Guide pages 56–59

Learner's Book pages 71–72

Double bubble template

Objective

Learners will revise:

- when balanced forces are acting on an object, it will not change its motion, shape or direction.

Science background information

If forces acting on an object are balanced, it means the forces are the same strength and acting in opposite directions. If balanced forces are acting on an object, it will not change its motion, shape or direction. If the object is still, it will stay still. If the object is moving, it will carry on moving at the same speed and in the same direction.

If the forces acting on an object are unbalanced, it means the force acting in one direction is greater than the force acting in the opposite direction. If unbalanced forces are acting on an object, that object will change its motion. If the object is still, unbalanced forces will make it start to move. If the object is moving, unbalanced forces will make it change speed, shape or direction.

Starter activity suggestions

It is useful to scaffold learners' use of the double bubble by constructing one with them related to something they are very familiar with, such as a children's story. By modelling how to make one together learners will then be better able to create one on their own. Encourage them to keep adding ideas to it as they learn more about the topic.

Revision approach background information

A double bubble is another type of thinking map. It is a useful way to check that learners have understood ideas. The double bubble helps learners to look at two different ideas and compare them for similarities and differences.

Activity notes and answers

Page 56
Activity 1: Answers

1 Child sitting
  a First arrow anchored on the seat of the chair pointing upwards and labelled 'support force of *chair* on the *child*'.
  b Second arrow anchored from the centre of the child's chest and pointing downwards and labelled 'weight force from the pull of the *Earth* on the *child*'. Both arrows MUST be the same size.

2 Teddy on table
  a First arrow anchored on the top of the table pointing upwards and labelled 'support force of *table* on the *teddy*'.
  b Second arrow anchored from the centre of the teddy's body and pointing downwards and labelled 'weight force from the pull of the *Earth* on the *teddy*'. Both arrows MUST be the same size.

3 Child floating
  a First arrow anchored on the surface of the water underneath the back of the child pointing upwards and labelled 'push force of *water* on the *child*'.
  b Second arrow anchored from the centre of the child's body and pointing downwards and labelled 'weight force from the pull of the *Earth* on the *child*'. Both arrows MUST be the same size.

Page 57
Activity 2: Answers

1 a–b

A Balanced forces, as the speed of the runner remains constant and is not changing.

B Unbalanced forces, as the motorbike is slowing down to stop, so the speed is changing.

C Balanced forces, as the speed of the swimmer remains constant and is not changing.

D Balanced forces, as the speed of the skateboarder remains constant and is not changing.

E Unbalanced forces, as the dog is jumping up from rest and begins to move so its speed is changing.

F Unbalanced forces, as the cyclist will continue to get faster, so the speed is changing.

The aim of the questions and follow-on activities are for the learner to see the following pattern: when forces are balanced there is no change in speed. When forces are unbalanced, there is a change in speed.
2–4 Check that learners have applied their learning correctly to their suggestions.

Pages 58–59
Activity 3

Example answers:

Similarities: use force arrows to represent any forces; involve two objects interacting; can be contact or non-contact forces.

Differences: length of force arrows – same size in balanced, different sizes in unbalanced; motion – does not change in balanced, does change in unbalanced; stationary – can be stationary with balanced, cannot be stationary with unbalanced.

Cross curricular

Work as a whole class to remind learners about how to create a rhyme. Give them time to create and share rhymes about everyday events, for example, playing, their favourite food or sport. Learners can then make a rhyme about forces.

Further activities

Challenge learners to create a diary for their day, picking out what they have been doing and which actions were balanced and unbalanced. Ask them to share their diary entry with a partner to peer assess.

ICT links

As a home activity, ask learners to create a picture collage of balanced and unbalanced forces.

Mass and weight

Study Guide pages 59–61

Learner's Book pages 74–75

Objectives

Learners will revise:

- mass is measured in kilograms (kg), and weight is measured in newtons (N)
- gravity is a force. When gravity changes, the mass of an object stays the same but the weight changes.

Science background information

Forces are measured in newtons (N). The unit is named after the scientist and mathematician, Isaac Newton. The instrument used to measure force is a force meter, also called a newton meter. A force meter has a spring inside it, connected to a metal hook. When a force is applied to the hook, the spring stretches. The greater the force applied, the longer the spring stretches, and the greater the reading on the scale.

The terms **mass** and **weight** are often confused. An object's mass is a measure of the amount of matter it contains. The more matter an object contains, the greater its mass. Mass is measured in kilograms (kg) or grams (g). Weight is the downward force exerted by a mass that is being pulled by gravity. Because weight is a force, scientists measure it in newtons (N).

The mass of an object is always the same, wherever it is in the universe. However, its weight depends on where it is. This is because it is the gravitational pull acting on an object's mass that creates its weight, and the effect of gravity is different in different places. For example, on Earth, the gravitational pull of the Earth on an object is equal to approximately 10 newtons per kilogram of mass. Other bodies in the solar system, such as the Moon and the planets, have different masses to the Earth's, and exert gravitational pulls of different strengths. The greater the mass of a body, the greater the gravitational pull it exerts.

Starter activity suggestions

Learners can find the words 'mass' and 'weight' confusing because in everyday life people tend to use the word **weight** when referring to measuring the amount of a substance in grams or kilograms.

It is helpful to refer to the amount of substance of an object as **matter**, and say we measure the mass of the matter in kilograms and the force needed to lift the matter as weight. The following activities are to help learners spot the patterns on Earth between the amount of matter (mass) and the force needed to lift them. The pattern is that for every 1 kg a force of 10 N is needed to lift it. The actual value on Earth is 9.81 N per kilogram. However, for young learners 10 is a more suitable unit to use.

Activity notes and answers

Page 59
Activity 1: Answers

Object	Mass in kg	Weight in N
Book	0.1	1
Chair	3.5	**35**
School bag	5.2	**52**
Table	**8.4**	84
Pencil case	**0.3**	3
Teacher's cat	10	**100**

Pages 59–60
Activity 2: Answers

1

Object	Mass in kg	Weight in N		Correct answer
Apple	0.02	0.02	✗	0.2 N
Shoe	1	100	✗	10 N
Bag of potatoes	2	200	✗	20 N
Rucksack	20	200	✔	
Skateboard	5	500	✗	50 N
Soccer ball	0.43	0.43	✗	4.3 N

2 They are either multiplying the mass number by 100 instead of 10, or not changing it if it is a decimal number.

3 All mass numbers need to be multiplied by 10 to find the weight.

Pages 60–61
Activity 3: Answers

1

Object	Mass in kg	Weight in N on Earth	Weight in N on Moon	Weight in N on Jupiter
Book	0.1	1	0.6	2.5
Chair	3.5	35	21	87.5
School bag	5.2	**52**	**31.2**	**130**
Table	**8.4**	84	**50.4**	**210**
Pencil case	**0.3**	3	**1.8**	**7.5**
Teacher's cat	10	**100**	**60**	**250**

2 Jupiter is the hardest place as the weight of the cat would be biggest there, meaning you would need to use a bigger pull force to lift it up.

3 The Moon would be where you could jump the highest. The pull of the Moon on you is less than the Earth or Jupiter, so your push force would make you go up higher.

Cross curricular

Challenge learners to imagine they are an astronaut walking on the Moon. They should explain via a video link what it feels like and the scientific reason for the difference between walking on the Moon and Earth. Ask learners to share their work and peer assess.

Further activities

Ask learners to bring three items from home to add to their table in Activity 3.

ICT links

With the whole class, watch a video of the Moon landing in 1969 and astronauts walking on the Moon. This could be used to support the Cross Curricular activity.

Gases, floating and sinking

Study Guide pages 61–65

Learner's Book pages 76–79

Graph paper

Objectives

Learners will revise:

- everything has matter, including gases
- a fluid exerts an upward force, called upthrust, on any object in it
- the mass and shape of an object can affect if it floats or sinks
- objects weigh less in water than they do in air.

Science background information

Gases are made up of particles and when they interact with another object, a force is exerted. It may be conceptually difficult for learners to perceive that gases are made up of mass (because of particles) because gases are often transparent, which learners may falsely believe means they are not made of any matter. Learners may also find it difficult to comprehend that the atmosphere is exerting a force on us constantly, as we have become de-sensitised and do not feel it.

Trying to find the mass of a gas is a challenging activity. It is not possible to calculate the mass of a gas using normal scales, as the gas is spreading out in all directions within a gas atmosphere. It would be like trying to use normal scales to find the mass of a bag of water while submerged in the sea; it is simply not possible. The only way to find the mass of a gas is to use a beam balance (lever balance).

A fluid exerts an upward force, called **upthrust**, on any object immersed in it. The upthrust is equal to the weight of the fluid displaced by the object. This depends not only on the volume of the object and how submerged the object is, but also on the viscosity (thickness) of the fluid. The more viscous the fluid, the greater the upthrust it exerts. The lower an object floats in a fluid the greater the upthrust exerted on it, until the object is fully submerged. Once an object is fully submerged, the force of upthrust can no longer increase. Water is more viscous than air, so it exerts a greater upthrust. It is the downward pull of gravity that gives a mass its weight, and upthrust acts in the opposite direction. As a result, objects weigh less in water than they do in air.

Starter activity suggestions

To support learners with visualising this difficult concept, it may be worth blowing up balloons with air and asking them what they think will happen when the balloon is released. Many will say it will float, as they will have seen helium balloons used at celebrations. However, when the balloon is released, it will sink. Learners can then have good discussions about why this is the case. The idea being that you blow more air into the volume of the balloon than there is in the same volume of air outside the balloon. It is therefore heavier and sinks. Learners can also add more key word cards related to these ideas.

Activity notes and answers

Pages 61–62
Activity 1: Answers

1 **a** Float up – the balloon of nitrogen with a mass of 14 g/mol is less than the mass of dry air, 29 g/mol, so it will rise in the air.
 b Sink to the ground – a balloon of oxygen with a mass of 32 g/mol is more than the mass of dry air, 29 g/mol, so it will sink in the air.
 c Sink to the ground – a balloon of carbon dioxide with a mass of 44 g/mol is more than the mass of dry air, 29 g/mol, so it will sink in the air.
 d Float up – the balloon of helium with a mass of 4 g/mol is less than the mass of dry air, 29 g/mol, so it will rise in the air.

2 If the mass of the gas is less than the mass of dry air, it will float up in the air.
 If the mass of the gas is more than the mass of dry air, it will sink to the ground in the air.

Pages 62–63
Activity 2

Learners have many alternative ideas as to why objects float and sink. These include thinking that there is less of a gravitational pull on an object in water (gravity is less), rather than understanding that the weight of the object will not have changed, instead the water is pushing up on the object which means less force is needed to lift it. This activity is designed to see if learners do in fact know the weight arrow does not change, and instead there is an upward support force arrow from the water, which makes it easier for the person to move the surfboard on the water. Once the person lifts the surfboard off the top of the water it will feel heavier as they will need to use a greater lift force as the push of the water no longer exists as the contact between the water and the object has been removed.

Answers

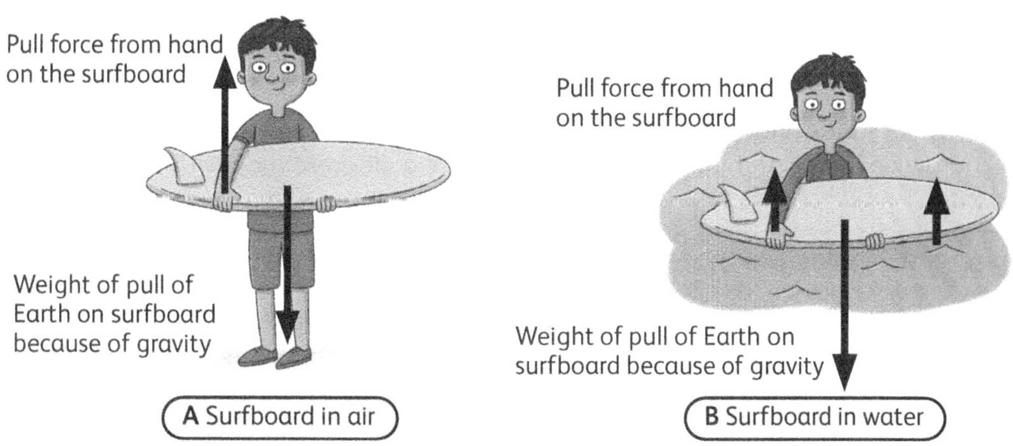

1 The image A of the child holding the surfboard in air is correct. Image B needs to change as the weight arrow is the same size as in image A as the weight of the surfboard has not changed. However, what is missing is an arrow anchored from under the surface of the surfboard where the water touches it. It should be pointing up and be labelled 'upthrust force of the *water* on the *surfboard*'. The size of the upthrust arrow should be the same size as the downward weight arrow. The upward arrow from the pull of the child's hands should be removed.

2 The surfboard floats because the weight force of the surfboard is balanced by the upthrust of the water on the board. The child therefore does not have to use as big a force to move the surfboard. However, as soon as the child lifts the surfboard above the water it will feel heavier again as they will have to supply the upwards force that the upthrust from the water had been giving.

3 They have not understood that the weight of the object on Earth does not change if the mass of the object stays the same.

Pages 63–64
Activity 3: Answers

1 Check that learners have drawn the force diagrams correctly.

2 **A** Sink, the upthrust force is less than the weight of the tanker.
  B Sink, the upthrust force is less than the weight of the child.
  C Float, the forces are balanced, the same size.
  D Float, the forces are balanced, the same size.

3 For the tanker (A) you could either make it out of a lighter material, take off some of the cargo it is transporting, or increase the surface area of the boat that is touching the water.
  For the curled-up child (B) they could spread themselves out like a star, so they are touching more of the surface of the water.

4 Objects float when the weight of the object is equal to the force of upthrust acting on them.

5 Objects sink when the weight of the object is greater than the force of upthrust acting on them.

Page 65
Activity 4

Supporting learners to make connections between key vocabulary helps them to remember the key words and it also helps them to develop their understanding about concepts.

Learners can be encouraged to use their key word cards to get their thoughts 'out of their head'. They could be asked to firstly just place cards they think go together in piles and be encouraged to display their key words and their thoughts in any way they want.

Once learners have thought about what they know about the key words, they can start to group them by identifying what they think the words have in common and create a concept map. Learners can then give the groups of words they have placed together a title.

Now learners are at the point where they can start mapping the words. Once a word and groups of words are placed, they can start to draw connecting lines between the different groups of words and between individual key words.

The learners can then write words on each of the connecting lines saying why they think the key words are connected.

Learners can either stick their final concept map together on a sheet of paper and keep it, or they could take a photo or draw an image of it and then try and produce a different map with similar or different connections on it.

Cross curricular

Using concept maps helps learners to not only recall words, but also supports them in developing their conceptual understanding of the key terms and how ideas link. This supports learners to develop literacy skills as they learn that it is important to not only know how to recognise and spell words but also that they can apply an understanding of a word in different contexts.

Further activities

Challenge learners to float some objects in water at home. They should draw a picture and annotate it with force arrows, or take a photograph and annotate it then bring back into class to share.

ICT links

Learners research which is the largest ship in use today.

Revision quiz

Answers

1 b

2 a

3 **a** newtons (N)

b Newton meter or force meter

c weight

d The bowling ball has a greater weight, or the apple has a smaller weight. Could accept the bowling ball has a greater mass or apple has a smaller mass.

e Apple 2 N; Bowling ball 8 N

4 **a** False

b True

5 **a** Steady speed – The motion that happens when balanced forces act on a moving object.

b Stationary – The motion that happens when balanced forces act on a still object.

c Speeding up – A motion that happens when forces acting on an object are unbalanced.

6 The four force arrow rules are that the arrows must:

- show the direction in which the force is acting. This is shown by the direction the arrow is pointing.
- show how strong each force is. The longer the arrow, the stronger the force.
- show where the force is acting from. The base of the arrow (the flat end) shows us where the force is acting from.
- be labelled with the name of the force, including the two objects that make up this force.

7 Any of the following: push, pull, support force, Normal force, upthrust, friction, air resistance, water resistance.

8 The **mass** of an object tells us how much matter there is in the object. This is not a force. **Mass** is not related (linked) to how the objects.

Weight is the force experienced by an object that is being pulled down by gravity.

The **mass** of an object will stay the same no matter where it goes in the universe.

The **weight** of an object will change depending on what different object interacts with it.

The **weight** changes depending on where in the universe it is.

9 **a** Graph title: Mass versus weight of objects on Earth

b Mass

c Weight

d

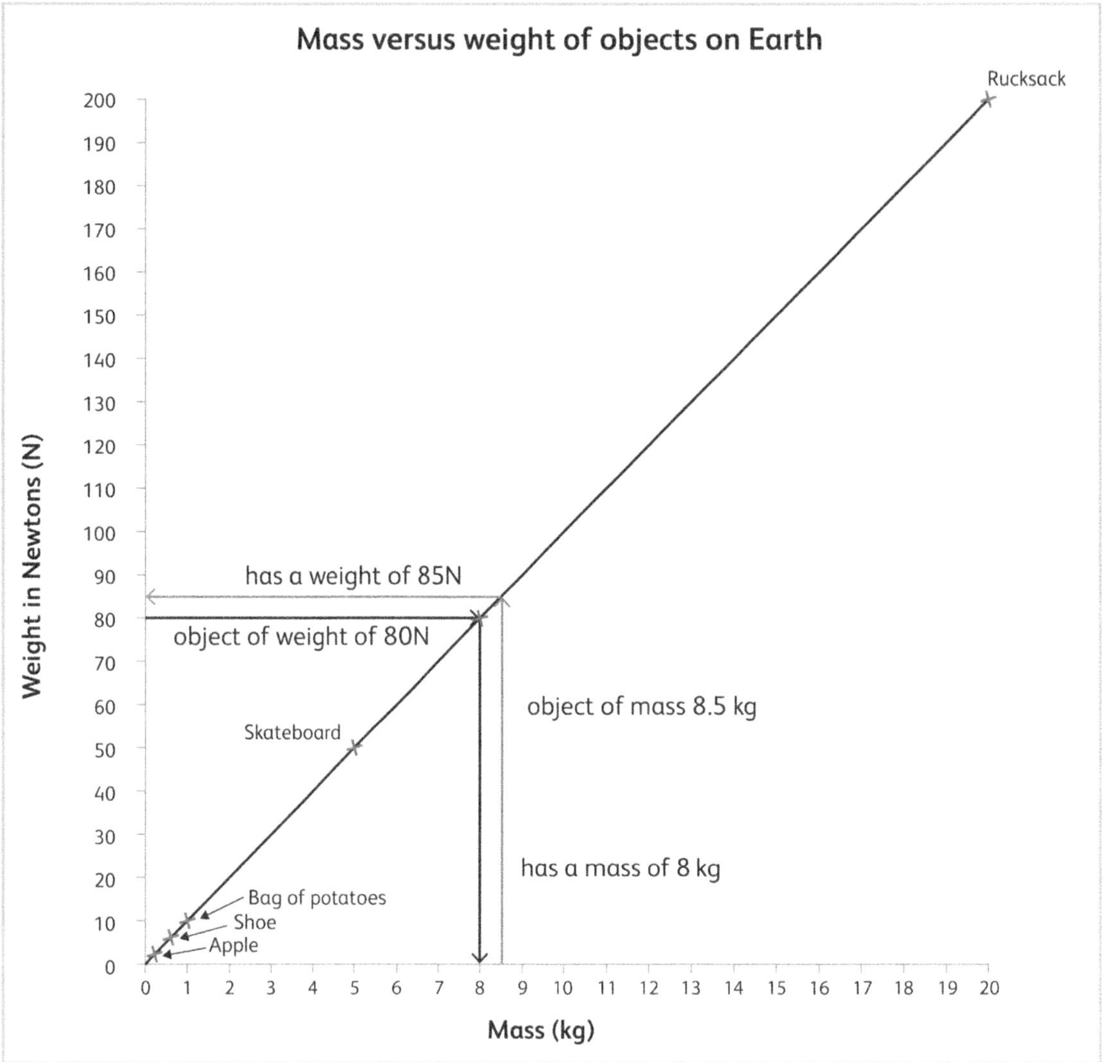

e Answer will be 8 kg

f 85 N

10 **a** True

b False

c True

d True

e False

Unit 6 Electrical circuits

Electrical circuits

Study Guide pages 68–71

Learner's Book pages 81–89

Objective

Learners will revise:

- electrical circuits can be drawn as a circuit diagram using symbols.

Science background information

A **circuit** is a complete loop, with no gaps, around which electrical energy can flow. A circuit shifts the electrical energy from a cell to make an electrical device work. If there is no cell, if one of the components is not connected correctly, or the circuit is broken, the electric current will not flow, and the electrical device will not work. As learners work through this unit, discuss the idea of electrical conductivity, that is, those materials that conduct electricity. Ask learners to say which parts of components are electrical conductors (allow electricity to flow through).

Starter activity suggestions

Learners are again asked to produce key word cards with images on them; these cards can then be used in several different ways.

Learners can be encouraged to work with others to play games with their cards to help them learn and recall the words, symbols and factual information. 'Snap' is an ideal game for learners to play with someone else that has also made key word cards for the topic.

If learners do not have access to another 'player', there are solitary games they can play. These include dealing out the cards picture side down and seeing if they can describe all the images before turning them over. They can deal out most of the cards and then try and work out which are missing. They could deal out all the cards and look at them then ask someone to remove one and shuffle them and see if they can work out which they took. They could create two sets of the images on the cards, deal them out and play a pairs game where they have to turn over two cards and see if they can match the pictures. They could play this game with another person too as a competition, and when a person matches a pair, they get to have another turn, and see who can win the most pairs.

Encourage learners to make additional cards as they encounter new key vocabulary throughout the unit and to use them in a variety of different ways to help their brain work to recall and retrieve information.

Revision approach background information

Key word cards are used again in this topic. The use of images in this topic is especially important as learners need to be able to recall and recognise the scientific symbols used to represent the different components in an electrical circuit. The key is to reinforce to learners that by regularly using their cards it will help them recall the key words, symbols and definitions and build their long-term memory.

Activity notes and answers

Page 68
Activity 1

Check that learners have used all of the words, and encourage them to be honest about the ones that they do not know or are unsure about. These are the words they need to focus on and learn.

Page 69
Activity 2: Answers

Correct circuit diagrams:

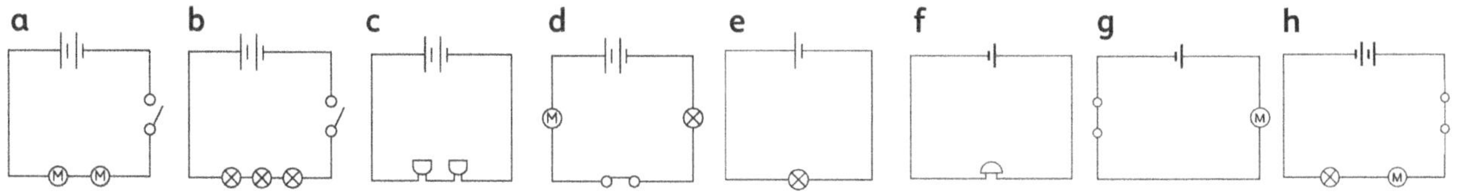

Page 70
Activity 3

The purpose of this activity is to check that learners can correctly associate the images of everyday components with the way they are represented scientifically.

The activity has been designed purposefully with errors to see if the learners are able to spot these.

It is important that you do NOT tell them how many of the images are correct as this will limit their engagement and thinking associated with the activity.

Answers

1 Learners A, E and F are correct representations.
2 Learners B, C and D have made mistakes; B, C and D have drawn a closed switch, B and D have drawn the cells the wrong way round.
3 Look carefully at whether switches are open and closed and the direction of the cells in the battery.

Page 71
Activity 4: Answers

Cartoon strips are a good way of getting learners to synthesise and summarise key points from the story of Alessandro Volta. To scaffold this for learners, you may wish to show them cartoon strips related to different stories, and get the learners to pull out what makes a 'good cartoon strip'. Criteria may include simple images, effective use of colour, limited use of vocabulary, key vocabulary being easy to spot, the story boxes being in chronological order, etc. Learners can then apply these success criteria to the Volta cartoon strip they produce. You could also help learners by supporting them in chunking the story into what they think are the eight key parts. Additionally further support can be given by identifying what the 10 key words are that they need to include.

1 Italy
2 He did not start talking until he was 4 years old.
3 The Voltaic Pile or electric battery.
4 He carried out experiments, questioned ideas and wrote to other scientists.
5 Not all scientists welcomed his ideas and disagreed with what Volta said was needed to originate the electrical flow in a circuit. It was because Volta was able to demonstrate through the use of experiments that alternating copper and zinc discs in his Voltaic Pile invention produced electricity in the form of sparks that the scientific community eventually agreed with him.
6 This can be any reasonable response that is qualified with evidence from the article, or the learners own additional research.

Cross curricular

Challenge learners to research Thomas Edison and his light bulb and to display the information creatively, for example, inside a picture of a light bulb.

Further activities

Challenge learners to find electrical appliances either in the classroom or at home and draw a picture of the circuit and then change it into a circuit diagram, for example, lamp, toaster (they can create their own symbol for heating element), mobile.

ICT links

Provide learners with an electricity game from the internet where they have to complete circuits.

Series and parallel circuits

Study Guide pages 72–77

Learner's Book pages 89–91

Double bubble template; Learning flower template; Graph paper

Objectives

Learners will revise:

- electrical circuits can be drawn as a circuit diagram using symbols
- circuits can be parallel or series
- series and parallel circuits can be compared to observe the effect on components such as lamps.

Science background information

In a circuit diagram, if there is only one electrical loop that learners can trace around with a finger, it is called a **series circuit**. In a series circuit, all the components are connected in the same loop.

When the number of components in a series circuit change, this will have an effect on the different components in the circuit.

If components are connected in more than one loop, this is called a **parallel circuit**.

Starter activity suggestions

It is possible to work with learners to create fruit batteries where there are two different metals used to make a clock LED light up. These work on the same principle as Volta's Voltaic Pile.

Learners can be provided with a range of already built circuits, comprising of both series and parallel circuits to get them familiar with how they look when built in real life.

Activity notes and answers

Page 72
Activity 1: Answers

Component	Observable change
Buzzer	Buzzer volume can decrease
Lamp	**Brightness of lamp can decrease**
Motor	**Speed of motor spin can decrease**

Pages 72–73
Activity 2: Answers

Circuit	Situation	Prediction	Reasons
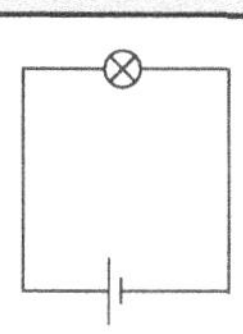	Another lamp is added to the circuit.	I think … Both bulbs will be dimmer.	Because … The two bulbs only have one cell now lighting both of them.

Circuit	Situation	Prediction	Reasons
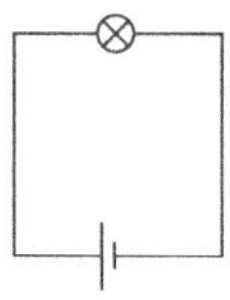	Another cell is added to the circuit.	I think … **Both bulbs will be brighter.**	Because … **There are more cells lighting the bulbs.**
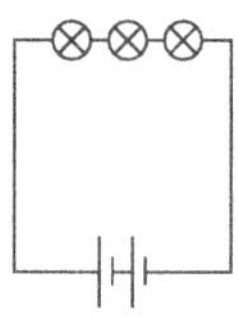	A lamp is removed from the circuit.	I think … **Both bulbs will be brighter.**	Because … **There are fewer bulbs in the circuit with the same number of cells.**
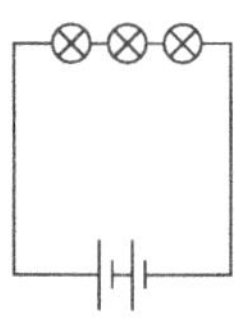	A cell is removed from the circuit.	I think … **The bulbs will be dimmer.**	Because … **There are fewer cells lighting the bulbs.**
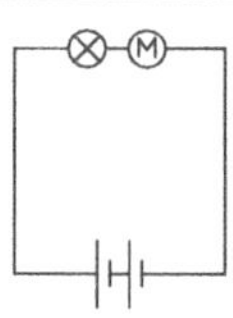	The motor is removed from the circuit.	I think … **The bulb will be brighter.**	Because … **There is now only the bulb in the circuit with the same number of cells.**
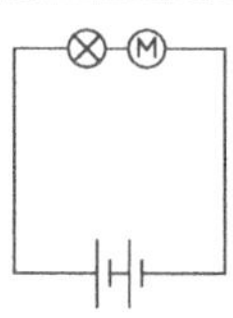	Another motor is added to the circuit.	I think … **The bulb will be dimmer and the motors will spin slower.**	Because … **There are more components in the circuit with the same number of cells.**
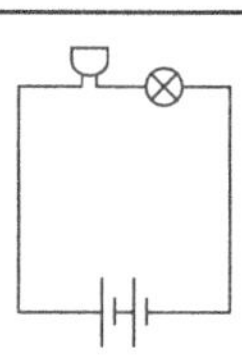	The buzzer is removed from the circuit.	I think … **The bulb will be brighter.**	Because … **There is now only the bulb in the circuit with the same number of cells.**
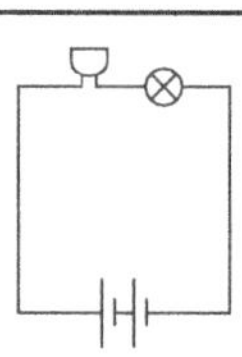	Another cell is added to the circuit.	I think … **The bulb will be brighter and the buzzer will be louder.**	Because … **There are more cells lighting the bulb and sounding the buzzer in the circuit.**

2 Example answers:

Adding more cells in a series circuit makes the bulbs brighter.

Removing cells in a series circuit makes the bulbs dimmer.

Adding more cells in a series circuit makes the buzzer louder.

Removing cells in a series circuit makes the buzzer quieter.

Adding more cells in a series circuit makes the motor spin faster.

Removing cells in a series circuit makes the motor spin slower.

Adding more bulbs in a series circuit makes the bulbs dimmer.

Adding more buzzers in a series circuit makes the buzzer quieter.

Adding more motors in a series circuit makes the motors spin slower.

Page 74

Activity 3: Answers

1 a A – 2, B – 2, C – 3, D – 3, E – 4, F – 4

b The same circuits are: A and B; C and D; E and F.

Pages 74–75
Activity 4: Answers

1

Circuit	Prediction	Reasons
A	I think … Both bulbs will be brighter	Because … Each loop in the parallel circuit now only has one bulb in it instead of two when it was in a series circuit.
B	I think … **The motors will spin faster**	Because … **Each loop in the parallel circuit now only has one motor in it instead of two when it was in a series circuit.**
C	I think … **The buzzers will sound louder**	Because … **Each loop in the parallel circuit now only has one buzzer in it instead of two when it was in a series circuit.**

2 Check learners' golden rule to make sure it works.

Pages 75–76
Activity 5: Answers

1 Voltage

2 The number of lemons

3 Possible answers for control variables include: size of lemons, juiciness (ripeness/age) of lemons, length of wires used between lemons, the shininess (cleanliness of the surface) of each type of coin used, same type of copper coins, same type of zinc coins.

4

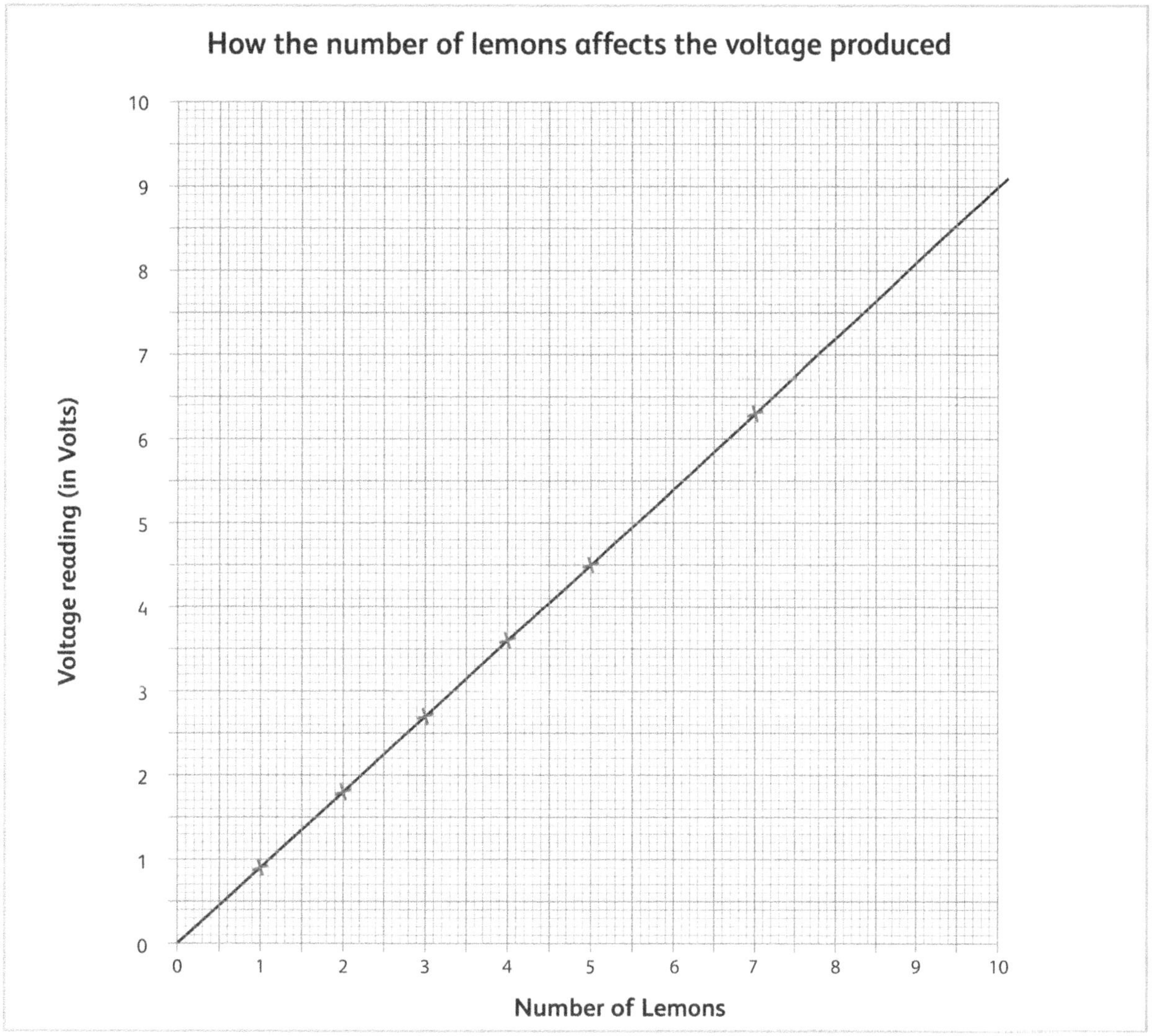

5 5.4 V, because each lemon is increasing the voltage reading by 0.9 volts.

6 **a** 3 lemons as they need 2.2 V and 2 lemons would only give them 1.8 V, which is not enough.

b 5 lemons as they need 4.2 V and 5 lemons would give them 4.5 V, which is enough to make it work.

c 8 lemons, this is because the LED lights are in a series circuit and together would need 6.4 V to work (2.2 + 4.2), 7 lemons give 6.3 V which is not enough so they would need an additional lemon to ensure they had enough voltage.

Pages 76–77
Activity 6

Provide learners with copies of the double bubble template. If you have a double bubble from a different topic available, use it to model how a double bubble works.

Example answers

Similarities: they need wires, they need a cell, they need a component, they have loops in the circuit, and they need complete circuits to work.

Differences: there is only one loop in a series circuit; there are multiple loops in a parallel circuit. Components are next to each other in a series circuit, components are in different loops in a parallel circuit. Adding lamps in a series circuit makes them dimmer; adding lamps in extra loops in a parallel circuit keeps them all bright. Adding motors in a series circuit makes them spin slower; adding motors in extra loops in a parallel circuit keeps them spinning fast. Adding buzzers in a series circuit makes them quieter; adding buzzers in extra loops in a parallel circuit keeps them loud.

Page 77
Activity 7

A learning flower is a memory aid that can help the learner's brain remember and organise ideas. It is a visual way of organising their learning, which some learners find useful when trying to revise a topic.

By summarising their learning in this way, it can help them recall important things they have covered in the topic and make links between the different ideas.

The learners need to use the flower to summarise the topic by:

- Giving it a title in the centre of the flower
- Adding key words on the front of the petals
- Adding definitions of the words on the back of the petals
- Noting their key learning on the stem
- Linking to what they already know on the roots
- Adding questions or things they do not fully understand on the leaves.

Cross curricular

Using learning flowers helps learners to not only recall words, it also supports them in developing their conceptual understanding of the key terms and how ideas link. This supports learners to develop literacy skills as they learn that it is important to not only know how to recognise and spell words but also that they can apply understanding of a word in different contexts.

Further activities

Give learners the opportunity to make fruit batteries either with lemons or other fruit, for example, limes, oranges or kiwi fruit to find out if they all work. They could do this as a home activity borrowing basic kit, and bringing their fruit battery to class to show others.

ICT links

Working in pairs, challenge learners to choose three internet sites on electricity for their age group, use the site and then write a review describing its positives and where it could be improved.

Revision quiz

Answers

1 All are correct

2 c

3 a

4 **a** True
   b True
   c False

5 **a** Switch – A component used to close and open gaps in an electrical circuit.
   b Series circuit – An electrical circuit that has only one loop.
   c Parallel circuit – An electrical circuit that has more than one loop.

6 **a** Cell 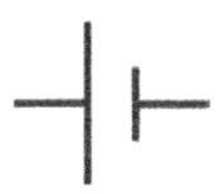
   b Open switch 
   c Closed switch 
   d Wire

7 **a** lamp
   b buzzer
   c motor

8 **a** Series
   b Parallel
   c Parallel
   d Series
   e Both
   f Series

9 **a** B
   b B
   c B

10 **a** False
   b True
   c False
   d True

Unit 7 Light, reflection and refraction

Reflection

Study Guide pages 80–82

Learner's Book pages 94–97

Revision hexagons template

Objectives

Learners will revise:
- light travels in straight lines
- when a ray of light is reflected from a plane mirror it changes direction.

Science background information

All waves (including light) are emitted from a source. They often interact with a medium as they travel and are then sensed by a detector. For example, the Sun (source) emits light waves that can reflect from a surface (interaction with an opaque material) and be seen by us with our eyes (detectors). Light can travel through transparent materials, partially travel through translucent materials but cannot travel through opaque materials. Most materials that light encounters are opaque.

One characteristic of the way light travels is that, as a beam of light moves outwards from its source, it spreads out to cover a wider area. Laser beams spread out much less than other types of light. A typical laser beam might spread out by only one metre when shone onto a surface one kilometre away. A misconception is that the light is dimming, this is not the case, it is just that the light from some sources spreads out.

Another characteristic of the way light travels is that it always travels in straight lines. A light ray can be represented as a straight line with an arrow. The arrow shows the direction in which the light ray is travelling. The light ray is a way of representing the journey that light waves undertake. It shows where the energy moves, in this way it is like the arrows used in food webs and food chains.

Starter activity suggestions

Learners can be given practise in visualising and then interpreting and drawing simple ray diagrams. One way to visualise the journey a light ray goes on is to drop a ball from a height onto a surface and on its rebound to catch it. This can be broken down to show learners there was a source, where the ball was dropped from, a surface the ball encountered and reflected from, and a detector, the hands that caught the ball. This could be linked to the Sun as a source of light, any opaque surface that reflects light and the eye that detects light.

Revision approach background information

Learners are again asked to produce key word cards with images on them, however, this time they are asked to use hexagons. Provide learners with a copy of the hexagons template to cut up.

The power of using a hexagon is that it has six sides. Learners need to place hexagons so they are touching, with the key activity being that the learners need to articulate why two sides are connected. Demonstrating how words connect will help build links for the learner and deepen their understanding.

It is good to challenge the learners to do this as a game in pairs. The hexagon cards are shared equally between the learners. Each person then takes it in turns to choose one of their hexagons and to lay it down to make connections with other cards, which they must articulate. Once all hexagons have been placed, learners can pick them all up, shuffle and re-deal them and play again. This way they will be learning about connections the other learner has also made.

Encourage learners to make additional cards as they encounter new key vocabulary throughout the unit, and to use them in a variety of different ways to help their brain work to recall and retrieve information.

Activity notes and answers

Page 80
Activity 1

Check how the hexagons are laid out, where you are unsure why learners have placed hexagons together, ask them to explain their ideas. This is a formative assessment opportunity where you can assess ideas and potential misconceptions.

Pages 81–82
Activity 2: Answers

1 Diffuse reflections: balloon, sponge, pen, clothes, yoghurt
Specular reflections: aluminium can, glass of water, ice cubes, spoon
2 They are dull in colour and have a rougher surface.
3 They are shiny, lighter in colour (almost white/luminous) and have a smoother surface.
4 Hexagon cards could use images of objects seen to help demonstrate the characteristics of diffuse and specular objects.

Cross curricular

Use the hexagon revision approach in another subject, for example, history, geography or Literacy when planning a story. Discuss with learners that this type of revision approach is not just for use in science but can be used in different curricular areas.

ICT links

Learners could create a picture collage of objects that show diffuse reflection and examples that show specular reflection.

Angles of incidence and angles of reflection

Study Guide pages 82–85

Learner's Book pages 94–97

Objectives

Learners will revise:

- light travels in straight lines
- when a ray of light is reflected from a plane mirror it changes direction
- reflected light ray bounces off the material at the same angle as the incoming light ray.

Science background information

A misconception that learners often hold is that only shiny objects reflect light. As they learn about reflection, it is important to stress that any object that is opaque must reflect light if we are able to see it. It is because the light bounces back (reflects) off the object that our eyes can detect it. Objects that are very good at reflecting light are those we see as the shiniest, however, objects that are dull/dark do still reflect light.

The law of reflection states that the reflected light ray bounces off the material at the same angle as the incoming light ray (the incidence ray); in other words, the angle of incidence equals the angle of reflection. When rays of light hit a smooth opaque surface, such as a calm lake, all the rays reflect in the same direction, producing (making) reflections. This is called **specular reflection**.

When rays of light reflect off most opaque surfaces, the rays are reflected in different directions because the surfaces are uneven and rough. Rough surfaces do not usually make clear images. This is called **diffuse reflection**. However, each individual ray of light still obeys the law of reflection; it is because the surface is uneven that the rays bounce off in many directions. This is known as **scattering of light**.

Starter activity suggestions

Make a dark den and ask learners to enter with an object which they think they will be able to see. Give learners a variety of objects to choose from including things that appear shiny and others that are darker or dull. Allow the learners to enter the den and describe what they see. Then ask the learners to re-enter the dark den with their chosen materials but this time tell them to take in a luminous object, such as a flashlight or string of LED lights and now describe what they see. Get learners to then explain using ray diagrams their dark den experiences.

Activity notes and answers

Pages 82–83
Activity 1: Answers

1 Pattern seeking

2 The learners' results are helping them work out the relationship (pattern) between the angle of incidence and the angle of reflection for each of their experiments.

3 Row 4 – Angle of incidence (in degrees) = 52 therefore the angle of reflection (in degrees) would also be 52, as the angle of incidence should be the same as the angle of reflection.

Row 7 – Angle of reflection (in degrees) = 81, therefore the angle of incidence (in degrees) should be 81, as the angle of incidence should be the same as the angle of reflection.

4 Experiment 3 and 6. Learners did not measure and shine the light at the correct angle of incidence, they may not have read the value off the protractor correctly, they may have moved the mirror while they were working. To improve the accuracy of their results they could have repeated the experiments until the average results were the same values.

5

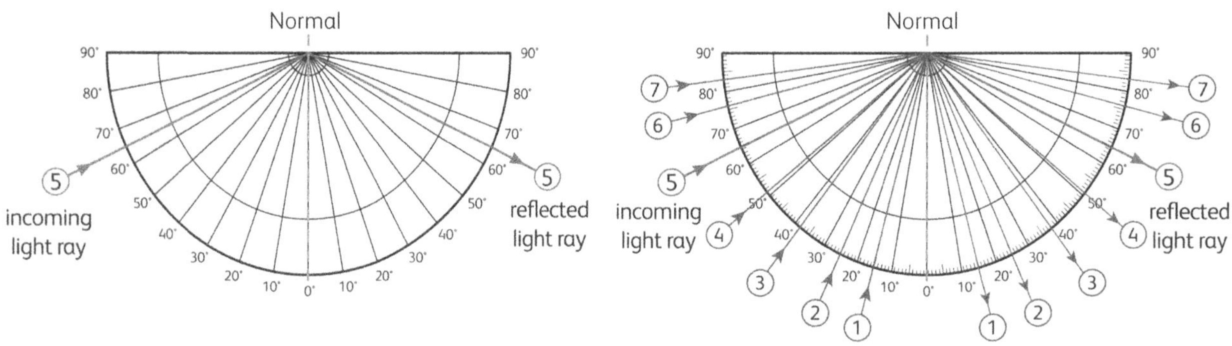

6 Yes there '*is a relationship (link) between the angle of incidence and the angle of reflection*'; they are the same as each other for every value. As one goes up so does the other.

Pages 84–85
Activity 2: Answers

1 Example answer: How does the angle of the mirrors affect the number of pencils observed?

2 Independent variable: the angle of the mirrors; dependent variable: the number of pencils observed.

3 The smaller the angle of the mirrors, the bigger the number of pencils seen, or the bigger the angle of the mirrors, the fewer number of pencils seen. Learners may notice that the angle of the mirrors multiplied by the number of pencils seen always equals 360.

4 Disagree, the learners are incorrect. The number of pencils would decrease, however, 120 multiplied by 2 equals 240. There would therefore be 3 pencils seen at 120° as 3 × 120 = 360.

5 The rule is that the angle of the mirrors multiplied by the number of pencils seen always equals 360.

6 Example answers: Dentist mirrors to easily inspect teeth; security mirrors in shops to help observe across a wide area; mirrors in vehicles to help drivers see clearly behind or at the side of the vehicle; magnifying mirrors to help people see things closer up when shaving or applying makeup; in flashlights to reflect the light and make a beam to light up areas better; periscope so that submariners can see boats and other potential targets or threats on the surface of the sea.

Cross curricular

As an art activity, learners could paint their own kaleidoscope patterns and display them, or draw a self-portrait and colour one half of their face in kaleidoscope patterns and colours.

Further activities

Using mirrors and tubes, learners make their own kaleidoscopes.

ICT links

Learners research different methods of making a kaleidoscope and make one at home, or research kaleidscope patterns to create a collage.

Refraction

Study Guide pages 85–94

Learner's Book pages 98–100

Double bubble template; Fishbone organiser template

Objectives

Learners will revise:

- light travels in straight lines
- when a ray of light travels through different mediums (for example, glass and water) it changes speed and can change direction, this is called refraction.

Science background information

Having learnt about what happens when light meets opaque objects, learners now learn about a different journey light can make and find out what happens when light meets transparent objects.

Light rays travel through transparent objects. When the light rays pass between the boundaries of transparent media, they slow down or speed up depending on the state of matter of the material. Light travels fastest in a vacuum, then a gas, then a liquid and slowest of all in a solid. Refraction occurs when the speed of the light wave changes. If the light hits the surface of the transparent medium at an angle, when the speed of the wave changes, the direction of the wave also changes. If the light hits the surface of the transparent medium travelling down the Normal (the perpendicular line 90 degrees to the surface), the speed of the wave changes but the direction of the wave does not change.

When light goes from a denser to a less dense material, it speeds up and changes direction and appears to bend away from the Normal. If it travels from a less dense to a denser material it slows down, changes direction and appears to bend towards the Normal.

Starter activity suggestions

It can be useful to show images to learners of objects seen through transparent media and observe how the object appears to change. You can hold up a glass of water and place an object behind it or in it, such as a pencil. The object often looks larger and distorted yet when you take it away from the glass of water it returns to its normal size and shape. The reason the object appears to alter is due to refraction and the effect the transparent medium has on the journey of the light ray.

Activity notes and answers

Pages 86–87
Activity 1: Answers

1 Fastest: air, water, glass/plastic, diamond: slowest.

2 Fastest is gas, then liquid and slowest is the solid.

3 and **4** Example answer:

In an order of best to weakest conclusions.

Learner 2 – the science is correct; they link ideas to the data given and provide reasons linked to the arrangement of particles in the different media.

Learner 4 – the science is correct, and they link ideas to the data given, they do not provide reasoning.

Learner 1 – this is weak because the statement is correct, however, they do not identify the pattern of increase or decrease in speed or relate it to the media, they also do not provide any reasoning, so the statement could be incorrect thinking if probed deeper.

Learner 3 – the pattern spotted is incorrect and therefore the rest of their reasoning is incorrect.

5 Sea water is more dense than fresh water because of the dissolved salt in the liquid. It is a liquid which is less dense than a solid so light will travel faster in it than in glass/plastic which are solids. If they give a number, accept an appropriate estimation, for example, between glass and water.

6 Check to see that learners have used three lines for their Haiku with the appropriate number of syllables in each. There should be 5 syllables in the first line, 7 in the second and 5 in the last, making 17 in total.

Pages 88–89
Activity 2: Answers

1 True:

A – goes from air to glass (less dense to more dense) and bends towards the Normal.

E – goes across the transparent boundary at 90 degrees so direction does not change.

False:

B – goes from water to glass (less dense to more dense) and bends away from the Normal, when it should bend towards the Normal.

C – goes from glass to water (more dense to less dense) and bends towards the Normal, when it should bend away from the Normal.

D – goes from air to water (less dense to more dense) and bends away from the Normal, when it should bend towards the Normal.

2 B, C, and D to be re-drawn as described above.

3 **a** Plastic to air – emerging ray bends away from the Normal.

b Diamond to water – emerging ray bends away from the Normal.

c Water to plastic – emerging ray bends towards the Normal.

d Water to air – emerging ray bends away from the Normal.

e Plastic to diamond – emerging ray bends towards the Normal.

f Diamond to plastic – emerging ray bends away from the Normal.

Page 90
Activity 3

1–5 All the devices shown: glasses; camera; binoculars; telescope; microscope; spyhole; magnifying glass; flashlight; movie projector use either a lens, or prism to work. Lens and prisms are transparent objects that refract the light enabling us to use the object.

The aim of this activity is to help learners engage in an approach that they can use to help them revise and build their long-term memory. By forcing the brain to recall information, it helps to make connections and embed ideas in the long-term memory.

Pages 91–92
Activity 4

Example answers:

1 Causes: Streetlamps, advertising boards, floodlights, our homes, etc.

2 Sub-effects: Birds flying into buildings; altered feeding, migrating and navigating patterns of plants and animals, for example, moths attracted to light bulbs and eaten by predators, insects attracted to headlights and killed by a moving vehicle, fireflies get confused by lights and cannot find mates to reproduce, newborn turtles get confused by light pollution and cannot use the light difference at the horizon to find the ocean, bats leaving lit areas to find darker parts of the world to live in, artificial light affecting human brains making it harder to switch off, fall asleep and get good sleep, artificial light may also affect humans making them more susceptible to illnesses such as diabetes, mood disorders and an increased risk of breast, prostate, and other cancers.

3 Solutions: Turning off lights at night, reducing the number of windows that are illuminated during the hours of darkness, looking at ways to reduce light pollution including requiring outdoor lighting to be shaded and not exceeding certain levels of brightness, closing curtains at night so the light does not escape, only lighting what needs to be lit and focusing security lights carefully.

4 a Provide learners with copies of the fishbone organiser template. Once learners have listed all the different causes, sub-causes and solutions they are then ready to create their own fishbone organiser diagram. The learners will only need to draw the number of 'bones' along the spine of their fish that correspond to the number of different causes they thought of. If you have a fishbone organiser from a different topic available, use it to model how a fishbone organiser works.

b At the top of each of the bones the learners write one of the causes, for example, streetlamps on one, advertising boards on another, etc.

c Learners then consider which of the sub-effects and solutions they can add to the bone they have labelled, for example, streetlamps has sub-effects of altering the mating of moths and also the migration of newborn turtles as they are both attracted to them. Solutions to the pollution of streetlamps is to turn of lights at night, shade the lights, reduce the brightness, etc.

Now encourage learners to use their fishbone diagram to help them learn and remember all the different ideas discussed in the article.

Page 93

Revision approach

If you have a double bubble from a different topic available, use it to model how a double bubble works. Provide learners with copies of the double bubble template.

Example answers:

1–2 Follow the instructions.

3 Similarities: they are both related to light; both need a light source to occur; both happen when light meets a material on its journey; both effects affect the journey of the light rays; both involve an angle of incidence.

4 Differences: reflection occurs when light meets opaque objects, refraction occurs when light meets transparent objects; light bounces off the surface in reflection, light passes through the medium in refraction; in reflection there is an angle of reflection, in refraction there is an angle of refraction; the angle of incidence equals the angle of reflection, the angle of incidence is different to the angle of refraction (unless the light travels down the Normal).

5–7 Encourage learners to regularly use their double bubble as a revision aid to support and help their brain remember the key ideas and the similarities and differences between reflection and refraction.

Learners can use the double bubble thinking map they have created to help them remember key ideas and see how the ideas of reflection and refraction are similar and different. Encourage learners to return to their double bubble over time to help their brain recall the information. If they find there are ideas they have forgotten, they can add a drawing at the appropriate place on their double bubble map to help them remember it next time, interestingly forgetting is a key part of remembering!

Page 94

Activity 5

Again, learners use their hexagons to help them consolidate the key language and ideas encountered in the topic and to make connections between them. If learners have not added any hexagons to their original cards, it is worth encouraging them to do so now. Getting learners to articulate their thinking of how cards connect is the key to helping them. Stress to learners there are no right or wrong answers with this, but there are right or wrong explanations. There are many ways they can link the hexagons; the key is that they can explain the links between touching hexagons. There are many ways the learners can use their hexagons to help them develop connections. A challenging activity is to get them to randomly pick a card and have this as the centre hexagon. They then pick six other cards and place them so that they make a circle around the original card. Learners must explain how the ideas connect where hexagons meet – for the outer cards there will be three connections for every card!

Cross curricular

Using hexagons can help learners to not only recall words, it also supports them in deepening their conceptual understanding of the key terms and how ideas link. This supports learners to develop literacy skills as they learn that it is important to not only know how to recognise and spell words but also that they can apply understanding of a word in different contexts.

Further activities

Create a poster to explain to other learners in the school and at home the problems of light pollution and how everyone can do something to reduce light pollution in their area.

ICT links

Challenge learners to research additional information about light pollution for their poster.

Revision quiz

Answers

1 aluminium can, glass of water, mirror, metal spoon, ice cubes, silver ring

2 b

3 a

4 a True
   b False
   c True

5 a Angle of incidence – The angle between the Normal and the incoming light ray.
   b Angle of reflection – The angle between the Normal and the reflected light ray.
   c Angle of refraction – The angle between the Normal and the refracted light ray.
   d Normal – A line drawn at 90° where light hits the surface.

6 a Reflected ray at 65°
   b Reflected ray at 20°
   c Incident ray at 80°
   d Incident ray at 70°

7 Example answers: spectacles (everyday glasses), camera, binoculars, telescope, microscope, spyhole, flashlight, movie projector.

8 a both
   b both
   c reflection
   d refraction
   e refraction
   f reflection
   g refraction

9 a Example answers: Material 1: air; Material 2: water. Reasons: light travels from less dense to more dense, slows down and bends towards the Normal.

- **b** Material 1: air; Material 2: glass. Reasons: light travels from less dense to more dense, slows down and bends towards the Normal.

  Material 1: air; Material 2: ice. Reasons: light travels from less dense to more dense, slows down and bends towards the Normal.

  Material 1: water; Material 2: glass. Reasons: light travels from less dense to more dense, slows down and bends towards the Normal.

  Material 1: water; Material 2: ice. Reasons: light travels from less dense to more dense, slows down and bends towards the Normal.

10 a False. It could speed up, it depends on the different material at the boundary.

- **b** False. It could speed up, it depends on the different material at the boundary.
- **c** True
- **d** True
- **e** False

11 a Example answers: Causes: streetlamps, advertising boards, floodlights and our homes.

- **b** Problems: Birds flying into buildings; altered feeding, migrating and navigating patterns of plants and animals, for example, moths attracted to light bulbs and eaten by predators, insects attracted to headlights and killed by the moving vehicle, fireflies get confused by lights and cannot find mates to reproduce, newborn turtles get confused by light pollution and cannot use the light difference at the horizon to find the ocean, bats leaving lit areas to find darker parts of the world to live in, artificial light affecting human brains making it harder to switch off, fall asleep and get good sleep, artificial light may also affect humans making them more susceptible to illnesses such as diabetes, mood disorders and an increased risk of breast, prostate, and other cancers.
- **c** Solutions: turning off lights at night, reducing the number of windows that are illuminated during the hours of darkness, looking at ways to reduce light pollution including requiring outdoor lighting to be shaded and not exceeding certain levels of brightness, closing curtains at night, so the light does not escape, only lighting what needs to be lit and focusing security lights carefully.

Unit 8 Rocks and soils

The layers of the Earth

Study Guide pages 97–101

Learner's Book pages 105–106

Frayer map template

Objective

Learners will revise:

- the Earth is made up of four main layers called the crust, the mantle, the outer core, and the inner core.

Science background information

Different kinds of rocks can be identified and classified by geologists (scientists who study rocks) according to their colours, textures and by how hard they are. Rocks are not universally hard and geologists use a 'scratch test' (more accurately, resistance to abrasion rather than hardness), the Mohs hardness scale published by Frederick Mohs in 1822 to classify rocks. Ideas related to porosity will be linked to ideas about hardness. **Porosity** is a measure of how much of a rock is open space. The open space is the gaps in the rocks, the holes. This space can be between grains or within cracks or cavities in the rock and the more gaps or holes there are, the more porous it is, allowing water to pass through it. **Permeability** links to porosity and is a measure of how easily a fluid, such as water, can move through a porous rock.

Rocks are materials and are found everywhere on Earth. The Earth's crust, its outer layer, is made up of rock, which is sometimes covered at the surface by soil or by water. Beneath the crust is the mantle. It forms 84 % of the Earth by volume, it is made of rocky material and magma (molten rock inside the Earth). While it is predominantly solid, it behaves like a viscous fluid because temperatures are close to the melting point. Deeper in the Earth are the liquid outer and solid inner core. Magma can escape through gaps in the crust of the Earth, through volcanoes. Magma on the outside of the Earth is named lava. Volcanoes can lie above land or beneath oceans (submarine volcanos).

Starter activity suggestions

There are a lot of key words that may be new to learners in this topic. To support learners with spelling and understanding these words, it may be useful to develop Frayer maps (see page 6 of the Introduction) for some of the key terms, which learners can refer to.

A Frayer map helps learners to identify and define unfamiliar concepts and vocabulary in a topic. As well as defining the word and characteristics associated with it, the Frayer map also provides examples and non-examples. This deeper analysis of the word helps to develop understanding. An example of a Frayer map can be seen on page 6. As well as using words on the Frayer map, pictures can be added into the four sections to support learning.

Some of the key terms in this topic to create Frayer maps include, igneous, sedimentary, metamorphic, fossils, rock cycle, solidification, erosion, sedimentation, burial, metamorphism, and weathering.

Revision approach background information

As there are so many words across this unit, learners are provided with the full list at the start and asked to identify those they already know and only create concept word cards for these. This is useful as an initial assessment activity as it can provide evidence of which learners have more understanding. Learners could then be grouped to help each other as they work through the unit.

As previously, once they have made their cards, learners need to organise them and place them onto a large sheet of paper. It is important that learners draw as many connecting lines as they can and write on the line the reasons why the cards are connected. As learners will be developing their understanding across the unit and repeating this activity at the end, they do not need to stick their cards down. It is therefore worth taking a photo of the concept map they produce. This is useful for the learners to refer to at the end of the unit so they can see how their ideas have developed.

Learners could do this activity in pairs and explain the connections they have made to each other. Learners could be paired following evidence gathered as to how many words they know, as a form of differentiating groups of learners.

Encourage learners to make additional cards as they encounter new key vocabulary throughout the unit.

Activity notes and answers

Page 98
Activity 1

Check concept maps, discuss with learners any words that they do not know. Look for common misconceptions across the class.

Pages 99–100
Activity 2: Answers

1

Object	Inner core	Outer core	Mantle	Crust
Indian Scotch egg	Egg yolk	Egg white	Meat	Breadcrumbs
Sliced apple	**Pip**	**Apple core**	**Apple pulp**	**Apple skin**
Sliced peach	**Seed**	**Peach stone**	**Peach pulp**	**Peach skin**
Wrapped sweet	**Nut**	**Caramel**	**Chocolate**	**Wrapper**
Sliced kiwi fruit	**White solid centre**	**Layer with black seeds in**	**Kiwi pulp**	**Kiwi skin**

2 Example answer: Best – Wrapped sweet; Indian Scotch egg; Apple; Peach; Kiwi – Weakest.

3 Reasons could include whether there is a liquid layer present; how easy it is to distinguish the four different layers of the Earth in the model; the size ratios of the different layers of the model.

4 Any reasonable idea where four layers exist, this could include other fruit such as cherries, plums, avocado, tomatoes, etc.

5–7 Give learners time to carry out this activity. Sharing with others will help to further embed learning and remember what they have learnt.

Pages 100–101

Activity 3: Answers

1 Correct order: B, C, A.

2

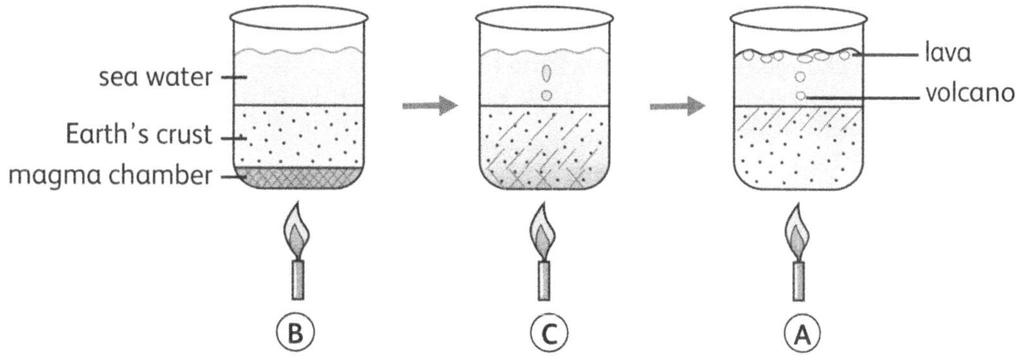

3 a Advantages: The model shows how liquids can turn into solids, like the lava turning into igneous rock; it shows that different liquids can solidify at different rates.

b Disadvantages: No crystals can be seen in any of the solidified wax; the model does not show the formation of intrusive igneous rocks, only the formation of extrusive igneous rocks.

4 That solid rock is formed at the top of the water when the liquid rock has cooled making solid areas of land that are surrounded by the sea.

Cross curricular

In geography, learners could map main areas of volcanic activity on the Earth's crust including submarine volcanoes.

Further activities

Challenge learners to research what kind of rock is local to where they live.

ICT links

Learners use, where appropriate, online dictionaries to check spellings and definitions of key words.

Types of rocks

Study Guide pages 101–107

Learner's Book pages 107–110

Objectives

Learners will revise:

- rocks can be classified as metamorphic igneous, and sedimentary
- the rock cycle explains how one type of rock can be changed into another over time
- weathering and erosion are part of the rock cycle.

Science background information

There are three major types of rocks on Earth, and they have different characteristics related to how they are formed. Each type of rock can be changed into the other types of rock.

Igneous rocks are formed when magma or lava is cooled, a process known as solidification, which results in crystals forming in the rock. Extrusive igneous rocks form outside the Earth's crust and cool quickly resulting in small crystals. Intrusive igneous rocks form inside the Earth's crust and cool slowly resulting in larger crystals because they have longer to develop. Igneous rocks are hard rocks.

Sedimentary rocks form from layers of small rock particles (sediments) washed into oceans. Sedimentation occurs as particles settle at the bottom of the ocean. Burial of the sediments occurs over millions of years and particles are squashed under pressure to form firm rock. Sedimentary rocks are softer and contain fossils.

Finally, **metamorphic rocks** are formed through heating and increased pressure acting on igneous and sedimentary rocks. The rocks are heated but do not become hot enough to melt and turn into magma. The heat and pressure are enough to change the properties of the rock.

Weathering and erosion are processes that break down rocks and the difference between them depends on where the process takes place.

Weathering is a process that breaks up rocks where they are located. Weathering happens by rain, wind, water, plants and animals.

When the smaller pieces of broken-down rock are moved from one place to another, this is called **erosion**. Erosion happens by wind, water and ice, and they move the broken-down rock to rivers, lakes or oceans.

Weathering often leads to erosion by causing rocks to break down into smaller pieces, which erosive forces can then move away.

So, if a rock is changed or broken but stays where it is, it is called *weathering*. If the pieces of weathered rock are moved away, it is called *erosion*.

Starter activity suggestions

Collect a range of different samples of rocks and challenge learners to interact with them to see how many ways they can classify them. It is also useful to take learners on a learning walk around their local school environment to see all of the different places that rocks and materials, such as concrete, that are made from rocks are used, i.e. in buildings or decorations such as monuments/statues, etc. Note that some learners hold the misconception that concrete is a rock. Learners can then research to find out the names of the different rocks and materials containing/made from rocks that are used. They can also see what qualities are useful for each of the uses of the rock and rock materials, for example, whether they are hard, smooth, non-porous, last a long time, etc.

Activity notes and answers

Page 102
Activity 1
Example answers:

1 and **2** Incorrect statements include:

Green circle – Lava is liquid rock outside the Earth's crust; magma is liquid rock inside the Earth's crust.

Pink circle – Intrusive igneous rocks form inside the Earth's crust and extrusive igneous rocks form outside the Earth's crust.

Purple circle – Chalk is not an igneous rock.

Activity 2

1 Any suitable picture of magma escaping the Earth and solidifying to form extrusive igneous rocks. This could be underwater, from a volcano, on the Earth's surface or from a volcano on the Earth's surface.

2 Encourage learners to create new concept map word cards.

Page 103
Activity 3

1 and **2** Six sections could involve: 1 – Igneous rocks being broken down by wind, rain, or by plants and animals into sediments; 2 – Erosion as rock particles are moved by wind, water and ice into rivers, lakes or oceans; 3 – Sedimentation as the sediment settles in the water at the bottom; 4 – Layer of sediment builds up at the bottom of the water; 5 – Burial as layers of different coloured sediments build up over time; 6 – Over millions of years the particles get squashed together and form firm sedimentary rock.

Pages 104–105
Activity 4: Answers

1 Pattern seeking

2 The learners can see patterns between the size of rocks and where they are in the guttering after erosion has occurred.

3 To see if their results showed the same pattern.

4 Yes, the larger rocks remained at the top of the guttering and the smaller pieces were found at the bottom of the tank. Therefore, in answer to their question '*Which size of rock particles are the most easily eroded?*', learners could say the smaller size rock particles were most easily eroded.

5 Rocks that were less than 0.5 cm were the most easily eroded.

6 Example answer: They are lighter, and it is therefore easier for the water to transport them.

Pages 105–106
Activity 5: Answers

1 C – Igneous because the chocolate is melted to become liquid chocolate, this represents solid rock being melted to form magma/lava. This liquid chocolate is then left to solidify to form the new solid igneous chocolate (rock).

2 A – Sedimentary because the chocolate (rock) is weathered by the grater, layers are built up to represent sedimentation and burial, and pressure is applied to try and turn the sediments into solid rock.

3 B – Metamorphic because the sedimentary and igneous chocolate is added together, they are then heated, squashed, and mixed together. This represents how sedimentary and igneous rocks are heated and squashed as they go deeper into the Earth to change (metamorphism) to form a new different rock due to the heat and pressure.

4 Suggested order: C – A – B. Igneous rocks weather and are eroded to form sedimentary, which are buried and heated to form metamorphic. However, as this is a rock cycle learners could start at a different point and justify their order.

Pages 106–107
Activity 6

Example answers:

1 and **2** Igneous – key words: crust, mantle, magma, molten, lava, solidification, igneous.

Sedimentary – key words: sedimentary, particles, erosion, sediments, sedimentation, burial.

Metamorphic – key words: metamorphic, metamorphism.

3 Encourage learners to create new concept map key word cards for new words they have encountered.

Cross curricular

Show learners a geology map from the internet, preferably of the area that they live, to find out about local rocks.

Further activities

Learners having found out about local rocks could explain how they are formed and put it on the school website.

ICT links

Learners could create a section on rocks and soils sharing their learning on the school website.

Fossils and rock cycle

Study Guide pages 107–109

Learner's Book pages 111–117

Objective

Learners will revise:

- fossils are found in sedimentary rocks, there are cast fossils and trace fossils.

Science background information

Fossils are the remains of plants or creatures, such as dinosaurs, which are no longer alive on Earth. Fossils were formed millions of years ago when animals and plants died and sank to the bottom of lakes, lagoons, or seas. Sediments of silt and sand were then washed down on top of the dead animals. In animals, the soft parts rotted away, leaving bones. As the sediments were compressed, they thickened, and over time turned to stone. The bones of the animals also decayed but left holes or 'natural moulds' in the sedimentary rock. Over hundreds of thousands of years, the holes or cavities filled up with minerals, which in turn became stone-like fossils. Sometimes the movement of the Earth's crust exposes fossils. Palaeontologists (scientists who study fossils) dig them up.

Fossils that form from the dead remains of parts of plants and animals were covered quickly after death by sediments such as mud and sand. In time, the sediments hardened to form rock and the bodies inside the sediments formed fossils, called **cast fossils**.

Other fossils are not formed by the actual plant or animal body but by things they left behind, such as eggs, solid wastes (their droppings) and tracks. These fossils are known as **trace fossils**.

Igneous rocks form from molten rock, and rarely have fossils in them. Metamorphic rocks have been put under great pressure, heated, squashed, or stretched, and fossils do not usually survive these extreme conditions. Generally, it is only sedimentary rocks that contain fossils.

The continual process of one type of rock changing over many millennia is known as the **rock cycle**.

Starter activity suggestions

This is an area of science that learners are very enthusiastic about as it links to ideas about dinosaurs and bones of other living things that once lived on Earth. You could prepare trays of soils with hidden items related to specific times and people, which learners can find, identify, and decide who they belonged to. These could be different historical inhabitants of your region. You and the learners could dress up as explorers. Learners could be given different trays and shown how to carefully dig to find what items are hidden. Once items have been found learners can do additional research to find out what life was like back in those times for the living inhabitants.

Activity notes and answers

Pages 107–108
Activity 1: Answers

Learning this sequence in order is important so that learners understand how fossils are formed and the length of time this process takes.

1 Learner 3. 3A – The animal originally died and fell to the bottom of the water where mud or sand covered some of the animals. 2B – Over time more bits of rock is washed down by rivers and covered the rotting animals. In this picture, the bones of the fish can still be seen and layers building up on top of it. 1C – The hard parts of the fish have remained and are changing and there are layers of rock now building up above it as it moves deeper into the Earth. 4D – After a long time, the fish parts have changed into fossils, the land on Earth moves and fossils can rise to the surface where they are then discovered.

2 Cast fossil because it is formed from the dead remains of an animal which has been covered quickly after death by sediments and in time, the sediments hardened to form rock and the bodies inside the sediments formed fossils.

3 False – Igneous rocks are formed from molten rock and any fossils in them would be destroyed at the high temperatures needed to melt the rock.

4 Encourage learners to create new concept map key word cards for new words they have encountered.

Pages 108–109
Activity 2

1–10 This is a useful activity for learners to help build memory and to identify where key processes involved in the rock cycle occur.

The 11 key terms the learners need to learn are: volcano, water, wind, ice, erosion, sedimentation, sedimentary, metamorphic, melting, magma, igneous. Learners can start with any word and should be encouraged to remember them in the order written and to explain why they are listed in this order related to the rock cycle.

Cross curricular

Ask learners to think about thousands of years in the future and what people living in the future would find as fossils? Which animals and plants living in their area today would form the fossils of the future?

Further activities

Challenge learners to find out where the word 'fossil' and 'palaeontologist' come from (*fossilis*, which means obtained by digging).

ICT links

Challenge learners to use the internet to research interesting facts, for example, worlds largest/smallest/oldest fossil.

Soil

Study Guide pages 110–114

Learner's Book pages 118–122

Double bubble template

Objectives

Learners will revise:

- weathering and erosion are part of the rock cycle
- there are different types of soils, we can classify them as sand, clay and silt and loam. Soils can be changed.

Science background information

The Earth's crust, its outer layer, is made up of rock, which is sometimes covered at the surface by soil or by water. In the same way that rocks are worn away to reveal fossils, rocks are also 'weathered', worn away by wind, rain, the action of freezing and warming and the action of animals. Over time this breaks the rocks down into tiny pieces which, when mixed with organic matter (dead and dying vegetation and animal matter – humus), makes soil.

As previously discussed, weathering and erosion are processes learners can find difficult to distinguish between as both can involve wind, water and ice. The main difference is the location of the process, if a rock is changed or broken down but stays where it is, it is called weathering. If the pieces of weathered rock are moved away, it is called erosion.

Soil is made up of different layers; at the surface is the darkest layer made mainly from organic materials (the litter or remains of dead plants and animals). Bacteria in the soil feed on these remains and break them down to form a substance called **humus**. The thicker layer under the surface is a mixture of humus and rocky fragments and is called the **topsoil**. This is the layer where plants are grown. Below the topsoil is the **subsoil**, this layer is paler than the topsoil because it contains much less humus. Below the subsoil is a layer containing lumps of rock, this layer is called **bedrock**.

There are different types of soils, and they can be classified based on their clay, sand and organic content. **Sand** is made up of larger particles that we can see with the naked eye and feels gritty. **Silt** is made up of particles too small to see with the naked eye and feels silky. Silty soils are often found in places that have flooded and dried out again. **Clay** is made up of tiny particles that fit together very closely and feels sticky. Clay soils are not as easy to squeeze together as the other types of soil.

Many soils are made up of a mixture of different soil types. **Loam** has a mixture of 40 % sand, 40 % silt and 20 % clay. It also has large amounts of organic materials, called humus. This gives loam soil special properties, making it the ideal soil for gardening and farming.

Starter activity suggestions

It may be useful to bring in examples of the different types of soil: sand, silt, clay and loam. Learners can then feel the different soils, examine them using a magnifying glass, add water to them, and squeeze them to look for differences between them. Learners can be encouraged to use the science words to describe how they look and feel, for example, gritty, silky, sticky. Seeing and engaging with the different types of soil is more likely to help them remember and recall the different names and their differing properties.

Activity notes and answers

Page 110
Activity 1: Answers

1 a Weathering
  b Erosion
  c Weathering
  d Weathering
  e Both

- **f** Weathering
- **g** Weathering
- **h** Weathering

2 Support learners to come up with signals to demonstrate breaking down in one place (weathering) and movement of rock (erosion).

Page 111

Activity 2: Answers

1. Bedrock layer – Chocolate and butterscotch chips. These are the largest bits of food which represent the larger pieces of rock found in bedrock.
2. Subsoil layer – Chocolate mousse. This layer contains a thick, dense substance and contains the same material as the layer below in a smaller form.
3. Topsoil layer – Crushed biscuits. This layer contains fine broken-down grains like soil.
4. Organic material layer – Green coloured shredded coconut and gummy worms. The green represents plants, and the worms show there are living animals too in this layer.

Page 112

Activity 3

Example answers:

Fact file – Sand

Sand particles are larger, and we can see them with the naked eye, they range from 0.05 to 2.0 mm.

The movement of rocks and minerals up and down a beach, rubbing against each other over thousands of years, results in sand. Most beaches are made from quartz sands. However, beaches may also contain other substances like volcanic glass, olivine, magnetic minerals, garnet, basalt, man-made particles like glass, organism shells and other living organisms such as coral skeletons, bivalve molluscs, algae, sea urchins, sponge spicule, and barnacles. May result in powder-fine grains of sand.

Beaches throughout the world and deserts.

Examples of uses of sand include cement, concrete, plastering, roofing, grouting, paint, to help defend buildings from flooding when it's in sandbags.

Fact file – Silt

Silt particles too small to see with the naked eye and range between 0.002 to 0.05 mm in diameter.

Sand and larger rocks are physically weathered to form silt, this includes rock being eroded and worn away by water and ice.

Silt can be found where it is deposited by water in places, such as wetlands, lakes, harbours, along riverbanks, on flood plains, as river deltas, where rivers deposit silt as they empty into another body of water.

Silty soil is usually more fertile than other types of soil and is used for growing crops.

Fact file – Clay

Clay particles are smaller than 0.002 mm in diameter.

Clay minerals most commonly form by prolonged chemical weathering of silicate-bearing rocks. Most clay minerals form where rocks are in contact with water, air, or steam (geothermal activity).

Clay can be found when there has been weathering of boulders on a hillside, in sediments on sea or lake bottoms, in deeply buried sediments containing water, and rocks in contact with water heated by magma.

Examples of uses of clay include bricks, cooking pots, art objects, dishware, musical instruments such as the ocarina, in paper making, cement production, pottery, and chemical filtering.

Fact file – Loam

Loam is a mixture of clay, sand and silt, depending how much of those soils are present, they can be called sandy loam, silt loam, clay loam or silty clay loam. Loamy soil is ideal for many plants because it holds plenty of moisture but also drains well so plant roots do not get waterlogged.

Pages 113–114

Activity 4: Example answers

1 Discuss with learners what an infographic is, you could refer them to page 7 in the introductory section of their Study Guide. Ask them to share ideas about the content of their infographic, reminding them that an infographic contains facts and information, particularly data, in the form of graphs, tables, charts, etc. Learners should draft out their infographic before creating their final version, they could peer assess and check each other's draft, providing comments relating to any errors or suggestions for improvement.

2 a The Wood Wide Web is an underground network in the soil of plant roots and fungi that allows trees to share nutrients and to lock up carbon dioxide in the soil.

b Scientists think that the Wood Wide Web locks up carbon dioxide in the soil removing carbon from the atmosphere, which helps to slow down climate change.

c Scientists think the fungal network could capture and take out five billion tonnes of carbon dioxide from the air – although it could be more than three times higher.

d Humans are changing the composition of soil and fungal networks through:

i agricultural expansion and deforestation when they change forests and grasslands to farmed fields and pastures. Change ways of farming, plant more trees, cut fewer trees down.

ii the use of fertilisers and pesticides – these are used to kill pests and add nutrients to the soil, they can also kill off the fungi and damage the fungal networks. Use organic farming, fewer fertilisers, and stop using pesticides.

iii urbanisation – this happens when green areas in towns and cities are taken and used instead for buildings, for example, homes, factories. Plan where to build carefully, do not use green areas and make sure that buildings are not left unused.

Page 114

Activity 5

Again, learners use their concept map key word cards to produce their final definitive concept map. It may be worth allowing learners to work with a partner first to talk through where they can group and place cards, as well as the links they can articulate that join different words. This way they will be able to consolidate their ideas and listen to those of someone else which could broaden their learning further. Getting learners to articulate their thinking of how cards connect is the key to helping them. Stress to the learners there are no right or wrong answers with this, but there are right or wrong explanations. There are many ways they can link the concept map key word cards; the key is that they can explain the links between connected cards.

If learners have not completed making all of the key word cards, the list on page 98 of the Study Guide can be shown to them and they can check which cards they do and do not have and make any additional ones they need.

Once learners are ready, they can stick their cards together and write all the connections they can think of, a photograph of this could be taken for learners to refer to later or they can keep and use their map to support their revision.

Cross curricular

Using concept map key word cards will help learners to not only recall words, but it also supports them in deepening their conceptual understanding of the key terms and how ideas link. This supports learners in developing literacy skills as they learn that it is important to not only know how to recognise and spell words but also that they can apply understanding of a word in different contexts.

Further activities

Encourage learners to carry out research at home on the Wood Wide Web and add their findings to their infographic.

ICT links

Learners could use ICT to create their infographic.

Revision quiz

Answers

1 granite, obsidian, basalt, pumice

2 b

3 a

4 **a** False
b False
c True
d True
e True
f False
g False

5 **a** Igneous rocks – Formed when magma from a volcano cools and turns into a solid.
b Sedimentary rocks – Formed from particles of materials over thousands or millions of years.
c Metamorphic rocks – Formed from other rocks that have been heated and squashed in the Earth's crust.
d Magma – Liquid rock beneath the Earth's crust.

6 **a** Solidification – A process that happens when a molten substance turns into a solid.
b Sedimentation – The process of particles settling at the bottom of a liquid.
c Burial – The covering over of rock particles by more and more layers.
d Metamorphism – A process of changing rocks through heat and/or pressure.

7 Example answers: wind, rain, ice in glaciers, moving water in rivers, oceans, onion skin, freeze-thaw, plants, animals, humans.

8 **a** weathering
b erosion
c weathering
d both
e weathering
f weathering
g weathering

9 **a** Algae
b Two – conifers and reptiles.
c Spiders
d Jurassic, birds and mammals – because they need flowering plants for food.

10 **a** False
b True
c False
d False
e True
f True
g True
h False

11 Agricultural expansion and deforestation; use of fertilisers and pesticides; urbanisation.

Unit 9 Earth and the Solar System

Earth and the solar system

Study Guide pages 118–121

Learner's Book pages 126–128

Objective

Learners will revise:

- the solar system is made up of the planets (including Earth), moons, asteroids, comets and meteoroids orbit around the Sun.

Science background information

Our planetary system is named the Solar System, this is because our Sun is named Sol, after the Latin word for Sun. Therefore, anything linked to the Sun, the word 'solar' is used, hence the term Solar System.

Our Solar System consists of the Sun, which is a star and then the planets Mercury, Venus, Earth, Mars, Jupiter, Saturn, Uranus, and Neptune, which orbit the Sun. All the planets (and moons, asteroids and comets) are kept in orbit by gravity between all the objects in our Solar System. This is why the planets keep the same distance and orbit around the Sun.

Our Solar System formed about 4.5 billion years ago. Scientists believe that it was formed from a huge swirling cloud of gas and dust. The force of gravity pulled parts of the cloud together. Most of the material formed the Sun, this material became so hot and dense it led to nuclear fusion, resulting in the light and heat of the Sun as we know it today.

The rest of the material orbited the Sun. Some of that material, the hard pieces, clumped together to form rocky planets (Mercury, Venus, Earth, and Mars). Further away from the Sun, where it is cooler, the gas and ice planets were formed (Saturn, Jupiter, Neptune, and Uranus). From this material dwarf planets, and large moons were also formed, this includes the Earth's own moon. Some of the rocky material did not form planets instead they became asteroids, comets, meteoroids, and small, irregular moons.

The Earth orbits the Sun, its orbit takes 365 days; as it does this the Earth rotates on its axis which is on an imaginary line from the North to the South Pole, one rotation takes about 24 hours. As the Earth rotates, one side faces the Sun, when this happens, we say it is in daylight, the other side facing away from the Sun is in shadow (darkness), known as night.

Revision approach background information

In this context, a rich picture poster is a strategy to support learners in bringing together what they already know on a topic, using pictures, diagrams, data and limited text. A rich picture poster helps to elicit learners' prior knowledge and understanding so that it can be shared and opens up what they know through discussion and peer assessment. A rich picture is an attempt to assemble everything that might be relevant to a complex situation.

Starter activity suggestions

Learners choose something that they are familiar with, for example, the Moon; ask them to use words, pictures, numbers and diagrams to create a big picture poster to show four things that they know. For example, learners might draw different phases of the Moon, say that the Moon is a satellite, draw a spaceship and write that humans landed on the Moon, they might even draw cheese and a sentence to say, 'Some people think incorrectly that the Moon is made from cheese'.

Discuss the idea of a rich picture poster and say that they have just created one and how it is used to share ideas about a topic. Get learners to go around the class and 'read' other groups' rich pictures and collect new ideas to add to their own.

Activity notes and answers

Page 118
Activity 1

Check learners' rich pictures, discussing with individuals or groups the content of their work, checking for misconceptions in their explanations of the words from the list.

Page 119
Activity 2

Check the rich pictures, discussing with individuals the mnemonic they have created, ensure the order and names of the planets are correct.

Pages 119–120
Activity 3: Answers

1 Check that learners have copied the information correctly. Give them time to share their interesting fact column with another learner, ask them to share the fact that they think is the most interesting from the other person's research.

2 **a** Jupiter because its diameter of 142 800 is greater than the other planets.
  b Approximately 886 million miles (1.4 billion km).
  c The table shows that Venus and Earth are closest together, they are 41.6 million miles apart.
  d They orbit in a counter-clockwise direction and the shape of their orbits is an ellipse.
  e In 2006 the International Astronomical Union (IAU) demoted Pluto from being the ninth planet from the Sun to one of five 'dwarf planets'.

Check that learners have researched and added an interesting fact and that they have completed the row asking which is a rocky or gas planet. Rocky planets: Mercury, Venus, Earth, and Mars. Gas planets: Jupiter, Saturn, Neptune.

Pages 120–121
Activity 4: Answers

1 Sun, Earth, axis, spin, day, night, 24 hours, lit, rotation, shadow (tilted is not a word that they necessarily need for this explanation, do challenge learners to think about whether all words need to be used).

2 The Earth spins/rotates. The Earth spins on its axis. It takes 24 hours for one rotation. When the side facing the Sun is in daylight, the other side is in shadow, this is night.

3 The Earth rotates on its axis, one rotation takes 24 hours. As the Earth rotates, one side faces the Sun, this side is lit by the Sun's light, when this happens, we say it is in daylight, the other side facing away from the Sun is in shadow, it is in darkness, we call this night.

Check that learners have added their completed explanation to their rich picture poster.

Cross curricular

Challenge learners to use geography skills and identify countries around the world and compare whether it is daytime or night-time at the same time as it is in their own country.

Further activities

Ask learners to decide which planet in our Solar System they are most interested in and to research that planet and share what they find out on their rich picture.

ICT links

Use the internet to research information. As a group of three, learners create a video of themselves modelling day and night with an explanation.

The Moon

Study Guide pages 121–123

Learner's Book pages 129–132

Phases of the Moon Calendar

Objectives

Learners will revise:
- the solar system is made up of the planets (including Earth), moons, asteroids, comets and meteoroids orbit around the Sun
- the Moon changes in appearance over its monthly cycle.

Science background information

There are over 200 moons in our Solar System, some planets like Saturn and Jupiter have many moons, 59 each but the Earth has only one Moon. Moons are also called natural satellites. The Moon takes approximately 27 days to orbit the Earth. The surface of the Moon is pitted with craters, has mountains and areas known as seas, although unlike on Earth they are not large bodies of water. Interestingly scientists have calculated that the Moon is moving away from the Earth, about 1 inch (2.54 cm) per year.

We can see the Moon from the Earth, at night the Moon shines bright in the night sky, however, unlike the Sun it does not produce its own light. We can see the Moon because it reflects light from the Sun.

The Moon is lit by the Sun, and it rotates as it revolves round the Earth, but we only see one side (face) of the Moon from Earth. As the Moon orbits the Earth, different parts of the Moon are in sunlight or darkness at different times. Looking from the Earth this makes it seem that the Moon is changing shape, because the Sun lights up different parts of its surface, we call these changing shapes **phases**. These phases have names, for example, Full Moon, Half Moon, Quarter Moon.

Revision approach background information

Making a model of the Moon, in this case, a card model of the phases, helps learners to visualise, feel and move (manipulate parts) to create the phases of the Moon. What is important is that the learners can describe what the different parts of the model represent, the black paper (when the Moon is not lit up) and the white card (Moon). They should also be challenged to describe what is happening and name the different phases of the Moon being represented.

Starter activity suggestions

Give learners a completed phases of the Moon calendar and ask them to describe how the phases change, explain what waxing and waning mean and identify the Moon phases, for example, Full Moon, Quarter Moon. Check that they recognise that this is a cycle that is repeated over and over.

Activity notes and answers

Page 121
Activity 1: Answers

1 Neil Armstrong

2 The Moon is the Earth's satellite.

A satellite is an object that orbits a planet. Star is the Sun and the Solar System is all of the planets that orbit our Sun.

3 27 days

4 We can see the Moon from the Earth because it is illuminated (lit) by light from the Sun.

5 b People on different parts of the Earth see the same side of the Moon.

Pages 122–123
Activity 2: Answers

1 Check that learners have followed the instructions correctly.

2 The shape of the circle – the Moon changes.
Check that learners are able to show Full Moon, Half Moon, New Moon, Quarter Moon.

3 a Pull the card out to make the Moon wax (get bigger).
b Push the card in to make the Moon wane (get smaller).

Page 123
Activity 3: Answers

1 Accept reasonable answers. It was a good model because I was able to show the different phases and they could be seen clearly.

2 Accept reasonable answers. The model is only 2 dimensional, it does not show the Sun or Earth.

Cross curricular

Research how the Moon is shown or represented in your culture/religion.

Further activities

Research and model a lunar eclipse. Use the internet to find the lunar calendar for eclipses in your country.

ICT links

Use the internet safely for researching information.

Science in context – The future of the Moon

Study Guide pages 123–124

Learner's Book pages 134–136

Objective

Learners will revise:

- the solar system is made up of the planets (including Earth), moons, asteroids, comets and meteoroids orbit around the Sun.

Science background information

Humans have already journeyed to the Moon and back, and scientists as well as entrepreneurs are thinking about and planning on how, in the future, humans could visit, live, and work on the Moon. There are still many problems to solve before this happens, for example:

- How to make sure space travel is safe?
- What will be used as an energy resource? Will they work as well on the Moon as they do on Earth?
- What kind of places will have to be built for people to live in? What are the problems? How will they get materials to the Moon?
- How will people grow food on the Moon, is there soil and water?
- How will people breathe on the Moon? How will they make sure there is enough oxygen?
- How dangerous is the Lunar dust? Is it poisonous to humans?
- How will people cope with the extreme temperatures on the Moon? (-414 to 253 degrees Fahrenheit). What are the temperatures like and why?

There are, of course, not just practical problems to overcome but an ethical debate which includes viewpoints such as:

- Humans should not live on the Moon because we cannot take care of our own planet and would likely not take care of the Moon.
- Some people say that we need a backup planet in case we need additional living space, resources.
- Humans can learn so much from space exploration and that includes the Moon.
- We need a new source of valuable resources that are running out on Earth that would help millions of people.

- Humans could ruin the Moon for future generations or even destroy kinds of life that humans have not discovered.
- Living on the Moon and taking its resources might lead to arguments between countries which could lead to war.

Starter activity suggestions

Ask learners to vote on the question, 'Should humans live on the Moon?' and record the results of the vote. Then ask learners to discuss PMIs (positives, minuses and interesting) related to this question. They could put their ideas on sticky notes and organise them under the three headings, then visit other groups to find out their ideas. Learners could return to their own area and add new sticky notes with ideas that they have collected from other groups. This could provide a useful beginning to the questions learners will be considering in following the activities.

Activity notes and answers

Page 124
Activity 1

1–2 Accept all appropriate answers.

Activity 2

1–5 Check learners' additions and changes, they could share their additions and changes with others explaining why they have made those amendments as well how their learning has changed in this topic. For example, what do they know now that they did not before, what do they understand better now, or still have problems with?

Cross curricular

Engage learners in a debate about humans living and working on the Moon. Debating ideas is important in science and is a way of encouraging learners to think through an idea and present evidence/an argument to justify what they are saying. It helps learners to realise that there are important topics that they should develop an understanding of and form a reasoned viewpoint based on facts. Part of science education is for learners to see themselves as global citizens and develop a willingness to question what is happening in the wider world.

It is also an opportunity for learners to apply and further develop their linguistic skills in relation to arguments with the use of phrases such as:

I believe …	The evidence suggests …	In my opinion …
My conclusion is …	I disagree …	I have mixed feelings …
Research suggests …	On the other hand, …	Do you agree that

Further activities

When learners have researched, considered and debated the arguments for and against humans living and working on the Moon, ask them to vote again. Has the result changed? Did they vote differently to the first time, why? What made them change their mind?

ICT links

Learners could video themselves presenting their argument for or against humans living and working on the Moon. By replaying the video in their group, they could engage in self and peer assessment on how they presented their argument. What was good about it and what could be improved and then repeat the process and consider the difference.

Revision quiz

Revision quiz tips

Refer learners back to their rich picture poster to read through what they know and have learnt this topic. Ask them to note anything that they are unsure about and to work out a way of learning/understanding the ideas or facts.

Answers

1 Pluto

2 No

3 c

4 **a** True
- **b** False
- **c** False
- **d** True
- **e** False
- **f** True
- **g** True

5 Lunar month – The time between the beginning and end of the Full and New Moon.

Phases of the Moon – The different shapes of the sunlit part of the Moon as seen from Earth.

The Solar System – The eight planets and moons that orbit our Sun.

6 Full Moon

New Moon

Half Moon

Also accept any other correct phases of the Moon.

7 **a** When the Moon is viewed from the Earth it appears to change <u>shape</u>.
- **b** These apparent changes in the shape of the Moon are called <u>phases</u>.
- **c** We can only see the <u>lit</u> side of the Moon.
- **d** At different points in the Moon's orbit around the <u>Earth</u>, different amounts of the Moon's sunlit side <u>face</u> the Earth.

8 **C** New Moon, **A** Full Moon, **B** Waning gibbous

Revision Test Answers

1

How microbes can enter the body	Body's defence mechanisms
Openings such as the mouth and nose allow microbes into the stomach	Mucous in the nose captures the microbes, we then sneeze and cough to blow them out of the body; for example, when we have a cold. Acid in the stomach can kill some microbes to stop people getting a stomach upset.
Different body surfaces such as the eyes, ears	Tears flush out microbes from the eyes and wax prevents microbes entering the ears.
Breaks, cuts or insect bites on the skin	The skin itself is a barrier to infection so if it is cut or grazed, it immediately begins to heal itself, often by forming a scab.

[6 = 2 marks for each correct row]

2 For example: burning, baking a cake, making toast, rusting, ripening of fruits, plant growing, human growing. [3]

3 a Correct [1]
  b Incorrect: These apparent changes in the shape of the Moon are called phases. [1]
  c Incorrect: We can only see the lit side of the Moon. [1]
  d Correct [1]

4 a living – dead [1]
  b mess – sediments [1]
  c softened – hardened [1]
  d blood – droppings [1]
  e track fossils – trace fossils [1]

5 a 1 lungs [1]
  2 heart [1]
  3 oxygen low blood [1]
  4 oxygen rich blood [1]
  5 rest of body [1]
  b Veins [1]
  c Arteries [1]

6 a For example: A food web shows how the living things in a habitat rely on one another for food. It shows many food chains, the connections between food chains and the different paths of energy. Food webs show the feeding relationships (links) between different living things and the energy transfer from one living thing to another. [4]
  b The insect and the chameleon should be swapped around. The food chain should show that energy is transferred from the plant to the insect to the chameleon to the fox. [2]

7 a For example: ash, smoke [1]
  b Wood [1]
  c Chemical change [1]

8 When we inhale (breathe in), air enters the lungs, and oxygen from that air moves to the blood. At the same time, carbon dioxide moves from the blood to the lungs and is exhaled (breathed out). [2]

9 a For example: In girls – breasts grow, start periods [1]
b For example: In boys – testicles get larger, develop Adam's apple [1]
c For example: In both – develop body hair, sweat more [1]

10 The relationship is that the angle of incidence equals (is the same as) the angle of reflection. [4]

11 Humans need a male and a female to **reproduce**. The male produces **sperm** and the female produces **eggs**. To create a new offspring (baby), the **sperm** must join with an egg, this is called **sexual reproduction**. [5]

12 The diaphragm moves upwards. [2]

13 For example: thermal insulators – plastic, wood, wool, air. Thermal conductors – metals [3]

14 a True [1]
b False [1]
c True [1]
d False [1]

15 For example: When **toxins accumulate** in an **animal**, **over tim**e this is called **bioaccumulation**. [5]

16 a – E solvent [1]
b – C solution [1]
c – D soluble [1]
d – B insoluble [1]
e – A suspension [1]

17 a For example: child sitting on a chair, bicycle not moving, cat sitting [1]
b For example: motorbike travelling at same (constant) speed, person jogging at same speed [1]

18 First arrow anchored on the top of the table pointing upwards and labelled 'support force of table on the teddy'. Second arrow anchored from the centre of the teddy's body and pointing downwards and labelled 'weight force from the pull of the Earth on the teddy'. Both arrows MUST be the same size. [4]

19 b [1]

20 c [1]

Total: 75 marks

Printed and bound by CPI Group (UK) Ltd, Croydon, CR0 4YY
10/03/2026
02069245-0001